Introduction to Scientific Thought

Newly Revised First Edition

EDITED BY
JOHN OAKES
Grossmont College

cognella
San Diego, CA

Bassim Hamadeh, CEO and Publisher
Michael Simpson, Vice President of Acquisitions
Jamie Giganti, Managing Editor
Jess Busch, Senior Graphic Designer
Seidy Cruz, Acquisitions Editor
Sarah Wheeler, Senior Project Editor
Alexa Lucido, Licensing Associate

First published in the United States of America in 2015 by Cognella, Inc.

Printed in the United States of America

ISBN: 978-1-63189-090-1 (pbk) / 978-1-63189-091-8 (br)

www.cognella.com 800.200.3908

Contents

Acknowledgement

By John Oakes

Special thanks to my colleague, Dr. Richard Albert. It has been my pleasure to teach Intro to Scientific Thought with him for many years. More than any other single person, he has helped me to think about the place of scientific thinking in the broader world. Many of the thoughts in this book that are found under my name reflect things I have learned from my good friend, Richard "Dick" Albert.

Section I
Science/Scientific Method

Introduction to Scientific Thought

By John Oakes

The path you are about to embark on will help you toward the goal of understanding and applying scientific thought to explain the wonders of the universe. This is not a science course: it is a course about science. A bit of science content will be slipped in here and there, but that is not the goal. The goal is for you to understand how scientists ask questions and how they answer them. You will be asked to look at the world from the point of view of the scientist.

The Scientific Worldview

If one desires to understand how another thinks—if you want to get into another's skin, a good first question to ask is "What does that person assume?" A person's working assumptions about the world form their worldview. So, we will begin our journey by asking what is the view of the world of the scientist. The answer is that when scientists looks at the world, they see order. They see a physical universe working according to laws that do not change, either in time or space. They see a world that can be understood by the observers—human beings. They see a world that can be fully described by abstract mathematics. These are the basic assumptions of science:

1. The universe is ordered—it is governed by physical laws that do not change...over time and space.
2. The laws governing the universe are obtainable and understandable by human beings.

3. The laws governing the physical world are fully describable by mathematics.

Carl Sagan described the scientific worldview as Cosmos. Cosmos means order. This worldview stands in stark contrast to that of most ancients, who saw the world as being a place of Chaos. The world was seen as a battleground, governed by a variety of gods, spirits, astrological forces, and fate. It was inherently unpredictable. The scientific worldview is a recent phenomenon on the time scale of human existence.

In a recent BBC radio show, a well-known chemist was put on a mock trial in front of a panel of jurists trying to probe the relationship between science and religion. The chemist was a philosophical naturalist who rejects the existence of any kind of supernatural reality. He insisted that the only things that are real are those which can be measured—that the only reality is the physical, and that there is no supernatural being. When asked how he knew this to be true, he pointed out the foolishness of certain religious beliefs as he perceived them. The jurists did not accept this nonanswer. They repeated the question a number of times, and he evaded answering. Finally, they insisted he answer the question: "How do you KNOW that the universe is governed by physical laws and that there is no supernatural reality?" His response: "Well, I guess I just believe it to be so."

This reveals something interesting about the fundamental presuppositions of science. They are accepted on faith. They cannot be proved. That the physical world is governed by unchanging laws is consistent with our experience as scientists and as humans, but it cannot be proved to be true. The assumptions of science also cannot be used to establish one way or another whether there is a higher reality. One can be an atheist, an agnostic, or one can hold to a religious worldview and still be a scientist. Are the working assumptions of science, as described above "True," with a capital T? The honest scientist will admit that they do not know if these assumptions are true in the absolute sense. Such an experiment cannot be done. What we can say is that from our experience, the fundamental assumptions of science are true for all practical purposes. When we attempt to describe the physical world using these assumptions, they work. In fact, they work spectacularly well. Whatever one thinks about the materialist assumption as a metaphysical position, we can all agree that the scientific worldview has had a powerful effect on the course of human history since it was accepted as a working model several hundred years ago.

So, where and when did this set of assumptions originate? The Greeks came close. Thales predicted a solar eclipse, showing in at least one important way that nature can be predicted using observation and mathematics. Pythagoras described the world using abstract mathematics. Aristotle viewed the physical world as having purpose (*teleos*). Yet the Greeks did not invent science. They had faith in deductive logic and reason, but failed to see the necessity of inductive processes, and were philosophically disposed not to do experiments. We will see that human reason alone, apart from experiment and induction, is singularly poor at discovering the underlying laws of nature.

The historical root of modern science is found in western Europe. It began with natural philosophers such as Roger Bacon (1214–1292). Arguably, Bacon was the first modern philosopher of science. He applied Christian theology to think about the natural world. His God was Creator: eternal, unchanging, and knowable. He concluded that the universe should be governed by unchanging, universal, ordered laws, which are knowable by human beings. In his great

treatise, *Opus Maius*, he proposed that knowledge of the natural world can be gained through "external experience, aided by instruments, made precise by mathematics." Here we see the first three elements of the traditional scientific "method," albeit in the "wrong" order. He proposed we should study the world through observation, experimentation, and forming mathematical hypotheses to describe those observations, confirmed by the experiments. Two of his three proposals were brand new. The Greeks observed, but they did not form mathematically precise hypotheses, and definitely did not design instruments to do experiments to test those theories. After Bacon, natural philosophy was to become an experimental endeavor.

The Scientific Revolution

Philosophers such as Bacon and William of Ockham laid the groundwork in the 13th and 14th centuries for what we now know as the Scientific Revolution, which occurred in the 16th and 17th centuries. The greatest figures in this revolution are Nikolai Copernicus, Johannes Kepler, Galileo Galilei, Robert Boyle, and Isaac Newton. We will spend considerable time in this course looking at their stories. Copernicus was perhaps the first modern "scientist," in that he made observations of the heavens and devised hypotheses designed to fit his data. He said "True assumptions must save the appearances." In other words, a hypothesis should only be proposed and accepted if it is in agreement with data. Reason alone cannot be trusted. His book *On the Revolution of Celestial Orbs* (1543) started a true revolution. His heliocentric (sun-centered) theory changed our view that we are at the center of the universe.

This revolution gained momentum almost a century after the career of Copernicus. Johannes Kepler applied mathematics to the precise astronomic observations of his mentor Tycho Brahe to create his three laws of planetary motion. Kepler, like Copernicus, had one foot in the modern scientific world and one in the medieval. According to Kepler, angels moved the planets in their precise elliptical. He made a living as an astrologer. With Kepler, the natural and the supernatural worlds were still blended.

Most historians of science will agree that the chief figure in the Scientific Revolution was Galileo. More than anyone else, he made science an endeavor performed in a laboratory, under carefully controlled conditions, using instruments specifically designed to do experiments. Galileo fought the battles with the religious establishment from which Copernicus refrained. He said in his letter to the Duchess Christina (1614) "The Bible was written to tell us how to go to heaven, not how the heavens go." He argued that the scientist should pursue the scientific evidence wherever it takes us. We should let nature speak for herself.

The philosophical debate during the Scientific Revolution was between the rationalists, represented by René Descartes (*Discourse on Method*) and the empiricists, represented by Francis Bacon (*Novum Organum*). *How* do we know? Is truth established by human reason, as Descartes argued ("I think, therefore I am"), or are observation and experience the arbiter of what is true, as Bacon argued. Descartes observed "The need to experiment is an expression of the failure of the ideal." If we read the writings of Galileo, Boyle, and Newton, Bacon won the argument, at least for a time. Experimentation was king during the 17th century. Theoreticians had to tread lightly. In the 21st century, the ideas of DesCartes and the preeminence of theory are making a comeback. Bacon was a visionary and a humanist. His vision was of human society transformed by technologies developed by scientists. He said that "Knowledge ought to bear fruit in work, that science ought to be applicable to industry, that men ought to organize themselves as a sacred duty to improve and transform the conditions of life." One thing we can say without fear of argument: Bacon's vision for applied science has proved true, although

the improvements brought by science have not come without a dark side.

The revolution started by Copernicus, Kepler, and Galileo was completed by Robert Boyle and Isaac Newton. Boyle helped to found the Royal Society, which became the leading forum for the presentation and evaluation of scientific ideas in the 17th through 19th centuries. In his *Sceptical Chymist* (1661), Boyle attacked the purely rationalist approach of Aristotle. He said that hypotheses are "the best we have, but capable of improvement." Here Boyle established what was not assumed before his time. It is fundamental to how scientists view their work today. All scientific conclusions are tentative. We do not discover "Truth." No scientific discovery is "true" in the metaphysical sense. The best we can hope for is to create explanations that are consistent with experimental evidence. Science is progressive. It changes over time. According to Boyle, it will always improve. Another of Boyle's statements demonstrates how those responsible for the Scientific Revolution broke with Greek rationalism. "We assent to experience, even when it seems contrary to reason." This is a classic statement of empiricism.

Perhaps it is a bit arbitrary, but it is convenient to think of Isaac Newton as completing the Scientific Revolution. His *Principia* (1689) was the greatest work of the Scientific Revolution—and perhaps the greatest in the history of science. In this great work, Newton described his discovery of calculus and his three laws of motion. Most important was his discovery of the universal law of gravity. When Newton derived Kepler's laws of planetary motion, the Scientific Revolution had its greatest victory. It is hard to exaggerate the importance of Newton's proposal that the laws of nature apply, not just to the earth, but to the entire universe. His "ah ha!" moment was not when he realized apples fall to the earth, but when he realized the reason the apple falls to the earth is also the reason the moon goes around the earth. Newton described his scientific method as follows: "The best and safest method of philosophizing seems to be, first to inquire diligently into the properties of things, and to establish those properties by experiences and then to proceed more slowly to hypothesis for the explanation of them. For the hypothesis should be employed only in explaining the properties of things, but should not be assumed in determining them."

The Scientific Method

The great victory of the Scientific Revolution was the development of the scientific method. You will spend a lot of time in this course learning about scientific methodology. So what, exactly, is the scientific method? Classically, it can be described as follows:

1. **Observation**. A scientist discovers a pattern: an apparent cause-and-effect relationship between variables.
2. **Hypothesis**. A hypothesis is a tentative explanation of the pattern discovered by observation. The hypothesis is a statement about a cause-and-effect relationship between an independent and a dependent variable.
3. **Experiment**. An experiment is a carefully controlled measurement of the relationship between the independent and dependent variable. The purpose of the experiment is to test the validity of the hypothesis.
4. **Verification**. The scientist(s) asks whether the data from the experiment supports the hypothesis, and whether it does so to the exclusion of other possible explanations.
5. **Predicted consequences**. The investigator seeks to broaden the implications of the initial experiment and hypothesis. If this is true, what else may also be true?
6. **Further experimentation and verification**.

With sufficient verification, if the concept of the hypothesis applies to a sufficiently broad range of phenomena, it can reach the status of a scientific theory. Contrary to common assumption, the principal distinction between a hypothesis and a theory is not the amount of experimental verification, but how broad a range of phenomena are explained.

Having described the classic scientific "method," it bears mentioning that there really is no scientific "method," if by such a "method" one means a set of steps which, if followed, will result in scientific discovery. Scientists make their discoveries through means not found in the traditionally described scientific method. Trial and error, serendipity, a search for beauty, intuition—and even sheer guessing—have led to important discoveries in science. Add to this the fact that no matter how many experiments one does, no hypothesis can be fully verified by experiment. In principle, inductive proof is unobtainable. If we hypothesize that all swans are white, no matter how many white swans we observe, we cannot prove that all swans are white. Perhaps there is a recessive gene for grey swans which is so rare in the population of swans that it is not expressed. Therefore, even if literally all swans were white, it would not "prove" by experiment that all swans are white.

In mathematical terms, because no measurement is infinitely precise, no scientific conclusion drawn from that measurement can be assumed to be "true." Isaac Newton's theory of gravity was true to one part in a million, and was in agreement with all known observations for over 200 years, yet it could not explain the precession of Mercury, and it was eventually replaced by Einstein's general theory of relativity. Is Einstein's theory "true?" Well, it is consistent with all the evidence related to gravity so far, but who knows if it will hold up to more precise measurements.

It is also worth noting that there is no logical connection between observation and hypothesis. Any single observation or set of observations may be explained by a number of hypotheses. There is no "correct" hypothesis. The means of choosing a scientific hypothesis is not necessarily rational. As Thomas Kuhn discovered, the acceptable range of hypotheses is determined, not by logic or common sense, but by one's working model. Kuhn called the working assumption in a particular area of science a paradigm. We will give considerable attention to Kuhn's theory of scientific revolutions and paradigms. The currently accepted scientific hypothesis is generally the one which explains the broadest range of observations and which has not yet been falsified. A "good" theory is one that is consistent with evidence, not one that is "true." Science is not about truth in the philosophical sense. We will see Karl Popper pointed out to scientists that the demarcation between science and nonscience is not consistency with observation and experiment, but falsifiability. An explanation which is, in principle, falsifiable by experiment but which thus far has not been falsified, is considered a successful theory. Dalton's atomic theory, which was the foundational theory on which chemistry was built, turned out to be false on two of its three principal claims.

In this course, we will consider a number of other questions about the nature of science. What are induction and deduction and their relationship to scientific inquiry? Is induction a reliable approach to discovering truth about nature? If no scientific statement is "true," then why is it that scientific methodology and thinking have been so fantastically successful at creating new technologies?

So what is science? It is not a search for truth. "Truth" will always be an elusive concept for science. Science is an unending story of the search for ever more successful and consistent explanations of patterns observed in nature. It is the quest for deeper cause-and-effect relationships in the physical world. It is a balanced use of inductive and deductive approaches to discover explanations of the underlying laws of the universe.

Ethics, Religion, and Science

If the scientific mindset is based on certain presuppositions, then it amounts to a worldview. Yet science is not done in a vacuum. In the real world, scientific thought must interact with other worldviews and philosophies. Many scientists accept the fundamental presupposition of science, but also hold to one of the major religious worldviews. Other scientists hold to a philosophical naturalism, denying the reality of any proposition other than that which is observable using the scientific method. Galileo said "For the Holy Bible and the phenomena of nature proceed alike from the divine Word, the former as the dictate of the Holy Spirit and the latter as the observant executor of God's commands." Einstein declared that "Science without religion is lame, religion without science is blind." On the other end of the spectrum, we have from Richard Dawkins, "In the universe of blind physical forces and genetic replication, some people are going to get hurt and other people are going to get lucky: and you won't find any rhyme or reason to it, nor any justice. The universe we observe has precisely the properties we should expect if there is at the bottom, no design, no purpose, no evil and no good. Nothing but blind, pitiless indifference. DNA neither knows nor cares. DNA just is, and we dance to its music." Or there is this from Richard Lewontin: "We exist as material beings in a material world, all of whose phenomena are the consequences of material relations among material entities. In a word, the public needs to accept materialism, which means that they must put God in the trash can of history where such myths belong."

It is not the purpose of this course to settle the question of what is the correct relationship between science and religion, but to challenge the student to think about what is the "territory" of the two. We will see that, as a rule, they are incommensurate—that they are radically different worldviews, with completely divergent vocabulary and methodologies. Science is quantitative, religion is qualitative. Religion offers authority. Science rejects authority. However, there is some overlap between the two. It is undeniable that both science and religion offer explanations of origins and the nature of human beings, which may or may not contradict. We will consider more than one perspective on how to think about the areas where they intersect.

Then there is the related question of ethics. Knowledge of the laws of nature gained by science is neither good nor evil. It is neither ethical nor unethical, but no one can deny that scientifically derived technologies raise major ethical questions. Science allows us to cure diseases, but it also gives us the ability to destroy all life on the planet. Some hail genetically modified crops as the solution to world hunger, while others see them as an environmental disaster waiting to happen. Maybe both are right. We will be considering what form the relationship between science, scientists, and society should take in view of the ethical issues raised by scientific discovery. We will also be taking a careful look at how the scientific community oversees ethics within its own ranks. Academic honesty is the hallmark of scientific inquiry. Without it, science is crippled by plagiarism, falsified data, or unwarranted, biased interpretation. How does science police its own honesty issues, and when might ethical violations of scientific integrity impact nonscientists?

Nonscience, Nonsense, and Pseudoscience

A major goal of this course is to train the student in skepticism. One thing we can say for sure is that there are a lot of urban legends, traditional beliefs, and outright fraud pushed on us as if they were scientific, when they are not. How is the nonscientist to wade through the myriad of voices and claims? What is the demarcation between science and pseudoscience?

In short, pseudoscience is any claim about the physical world which is not supported by well-documented, reproducible scientific evidence. By this definition, religious, philosophical, artistic—or any of a variety of other claims—are not pseudoscience. Our goal here is not to impugn the motivations of pseudoscientists. If the one offering unscientific "alternative" medical therapies is intent on committing fraud, if they are sincerely misled, or even if they are actually correct in their belief, but simply are putting forward a belief which is not yet supported by scientific evidence, this is not our concern here (although these are important questions!). The fact is that just because something is not yet "scientific" does not mean that it is untrue. In fact, we will consider some borderline cases such as continental drift, which appeared at one time to be pseudoscience but is now a major scientific paradigm. However, if we are being asked to believe that magnetic therapy is scientific when it is not, that is something we should be aware of. Some pseudoscience is fairly benign, but some is outright dangerous, as desperate people seek hope from unproved therapies. If nothing else, we may save ourselves a bundle of wasted money. Tens of billions of dollars are spent annually by believers in pseudoscientific claims. Our job is to shed some light on these claims.

Fortunately for us, to the trained eye, pseudoscience is generally easy to spot. A leopard cannot change his spots and pseudoscientific claims are almost without exception identifiable by traits we will call "marks of pseudoscience." The key is to learn the tricks of the trade. If it walks like a duck, it may or may not be a duck. However, if it walks like a duck, looks like a duck, smells like a duck, quacks like a duck and has the genetic material common to ducks, one can be assured it is a duck. This analogy will serve us quite well as we try to ferret truth from lie and real science from false scientific claims.

True or False: Test Your Knowledge of Scientific Methodology

Hopefully, reading the introduction has prepared you to answer many of these questions. [Adapted with permission from a similar unpublished list by Dr. Richard Albert, Grossmont College.]

1. Scientists usually expect an experiment to turn out in a certain way. _____
2. Science only produces tentative conclusions: ones that are likely to change. _____
3. Science has a more-or-less uniform way of conducting research we call the "scientific method." _____
4. Scientific theories are explanations; they are not facts. _____
5. When being scientific, one must have faith only in what is justified by empirical evidence. _____
6. Science is about facts, not about human interpretations of those facts. _____
7. To be scientific, one must conduct experiments. _____
8. Scientists manipulate their experiments in order to produce particular expected results. _____
9. Scientists prove conclusions that are definitive and final. _____
10. An experiment can prove a theory true. _____
11. Scientific explanations are based significantly on beliefs, assumptions, and nonobservable things. _____
12. Imagination and creativity are used in all stages of scientific investigation. _____
13. Scientific theories are ideas about how things work. _____
14. A scientific law is a theory that has been extensively tested and thoroughly confirmed. _____
15. The principal difference between a scientific hypothesis and theory is the amount of evidence supporting the idea. _____
16. Scientists' educational backgrounds, opinions, disciplinary focus, guiding assumptions, and philosophy significantly influences their interpretation of available empirical evidence. _____
17. A scientific theory will not change because it has been proven true. _____
18. A scientific law describes relationships among observable phenomena, but does not explain them. _____
19. Scientists invent explanations, models, or theoretical entities that are not observable. _____
20. Scientists construct theories in order to guide further research. _____

A History of Knowledge

By Charles Van Doren

Of all The kinds of knowledge that the West has given to the world, the most valuable is a method of acquiring new knowledge. Called "scientific method" it was invented by a series of European thinkers from about 1550 to 1700.

The genesis of scientific method goes back to the classical Greeks. Like all their gifts, it bears watching. But even though scientific method sometimes seems as dangerous as it is beneficial, we could no longer live without it.

So far in this book, when we have used the word *knowledge* we have usually meant what *anyone* could know. In medieval Latin "Knowledge" was *scientia*, and everyone could possess some or all of it. From the Latin comes our modern term *science*. But "science" no longer means the knowledge that anyone has or may have.

It does not mean a poet's knowledge, for instance, or a carpenter's, or even a philosopher's or a theologian's. Usually, it does not mean a mathematician's knowledge. "Science," today, is a special kind of knowledge possessed only by "scientists." Scientists are special people. They are not anybody.

The Meaning of Science

So much is probably obvious. Yet there are complexities in the meaning of "science" that are hard to unravel. Let us try using the world *science* in some sentences.

1. Science will never understand the secret of life.
2. Sooner or later, scientists will find a cure for AIDS.
3. Science and art have nothing in common.
4. I'm taking a science course, but I am also going to study some history.
5. Mathematics is the language of science.
6. Scientists are trying to determine if Shakespeare actually wrote all the plays that are ascribed to him.
7. Literary criticism isn't really scientific because it isn't predictive.
8. Most poets glaze over when they come upon a mathematical formula; most scientists glaze over when they come upon a poem.
9. Being bilingual doesn't mean you know anything about language.
10. I know the answer, but I can't explain it.

All of those sentences are "real" in the sense that they were taken from published sources and were written by respectable authors (Sentences 4, 9, and 10 were recorded from oral communications by respectable speakers). What do I mean by "respectable"? I mean that the authors or speakers were reasonably well educated and seriously meant what they said; that is, they thought that what they said was both comprehensible and true. Furthermore, all of the sentences are modern in the sense that they were composed within the last ten years. They clearly represent some kind of modern consensus about the meaning of the word *science* (which does not appear in the last two sentences, but is implied in both of them; that is, it is hidden or imbedded in the word know).

Let us examine a few of the sentences. The first one, for example: "Science will never understand the secret of life." Is this true? Manifestly scientists have recently, and in some cases not so recently, discovered many of the "secrets" of life, among them the structure and evolution of cells, the operation of the immune system, the role of DNA in genetics, and a great deal more. And we can expect scientists to go on studying life, and finding out its secrets. But there is something about the word secret in that sentence that makes the sentence both true and incontrovertible. By definition science is not able to understand the kind of secret that the secret of life is supposed to be, which by implication has something to do with an unfathomable mystery. Some other kind of knowledge is evidently required to solve *that* mystery, no matter how much knowledge scientists have about life, now or in the future.

Or take Sentence 5: "Mathematics is the language of science." This clearly proclaims that mathematics and science have a close relation, but it just as clearly proclaims that they are not the same thing. Scientists may *use* mathematics, but they do not *do* mathematics; and mathematicians can be just as ignorant of scientific methods and result as ordinary laymen are. Albert Einstein was a great theoretician but not a great mathematician; when he got into a fix he would go to his mathematician friends, who would invent the mathematics to get him out of it. But his friends, with all their skill, could never have come up with the theory of relativity.

At the same time, the sentence seems to say that mathematics is a different kind of language from French or Chinese, or from the language of body movements or musical notation. All of those are languages of a sort, but none o them could ever be called *the* language of science, although scientists might study any one of them.

Sentence 7, "Literary criticism isn't really scientific because it isn't predictive," is very curious. It is an old chestnut that science is not science unless it is predictive; that is, you do not really know something about the way nature works unless you can predict how it is going to work under this literary criticism (as, for example, the book review in the daily paper) is to tell you whether you will like (or be interested in) a book. Of course, the predictions are not certain. But not all experiments turn out the way you expect them to, either. Nor is the Judgment of the critic couched in mathematical formulas.

I would be the first to admit that literary criticism is not science, in the ordinary sense of the term. But I do not believe this is so because it fails to make predictions. Nevertheless, the sentence gets at a feeling we have about science, and contributes to the meaning of the word *science*.

Sentence 9, "Being bilingual doesn't mean you know anything about language," gets at another fundamental feeling that we have about science, whether or not we should have it. That is, it proclaims, by a wonderful indirection, that the kind of knowledge that anyone must have in order to do something consistently well, like speaking two languages, is not scientific knowledge. By implication, scientific knowledge, in itself, is not scientific knowledge. By implication, scientific knowledge, in itself, is not practical or useful. This sentence says nothing good about science. Most people would rather be bilingual than a scientific linguist. Bilingualism, in fact, is good for the brain (it makes it work better and faster), whereas knowing all about linguistics is of little use unless you want a job as a university teacher. The implication of the sentence is that often if not always, the knowledge that scientists have is specialized and relatively useless for ordinary persons.

However, Sentence 2, "Sooner or later, scientists will find a cure for AIDS," expresses our deep faith in science, our sense that we have to and can depend on science to solve the really hard, pressing, practical problems that we face. The sentence also suggests our sense that only scientists can be expected to find a cure for AIDS. Poets, carpenters, and philosophers, we are sure, will not find any such cure. Nor will an ordinary person, just by thinking about it, intuit a cure. This is one of the most widely held notions that go with the word.

In our scientific age, most teachers, hearing a student say Sentence 10, "I know the answer, but I can't explain it," would be tempted to respond, "If you can't explain it, then you don't know it!" and to give the student an F for presumption. Knowledge that cannot be framed and communicated, mathematically or otherwise, is not knowledge, in other words, and is certainly not scientific knowledge, which is felt to be (perhaps preeminently) public knowledge in the sense that it can and must be statable so that other scientists can test and validate it.

But this is to rule out as science, which once meant all kinds of knowledge, as we have seen, a vast panoply of human mental states and acts that do not have the kind of inherent certainty that scientific knowledge is supposed to possess. The best detectives always have hunches they cannot explain but that nevertheless turn out to be right, at least in fiction. Great athletes have an inexplicable and inexpressible genius when it comes to knowing where or how to run or throw the ball. Soldiers who survive may often do so because of their sixth sense about danger. And saints are more certain than any scientist about what God has told them, or about what they know about God in some other way.

However, we are not trying to prove the sentence wrong, and in fact it is not wrong, for it expresses something we feel about science, namely, that it cannot be exclusively intuitive, although intuition may be somehow involved in any important scientific discovery or breakthrough.

Finally, Sentence 3, "Science and art have nothing in common," reveals what is perhaps our deepest prejudice about science—and about art—at the same time that it is manifestly not true, at least on the surface. That is, science and art have many things in common, for example, in that both are activities involving some of the most capable men and women, that both science and art enlighten us and give us surcease from pain, that both are immensely difficult and require every ounce of effort and intelligence to succeed in them, that only human beings do them, and so forth.

But the sentence is true in another sense, which is also suggested by Sentence 8. We are pretty sure that scientists and artists, even if many of the things they do are similar—think of a metallurgist and a sculptor in metal—see what they do in different ways and do it for different reasons. It is their different viewpoint that tells us most about what "science" means and what "scientists" do.

Three Characteristics of Science

Science, then, in our common everyday sense of the word, is a human activity characterized by three things. First, science is practiced by special people with a specific view of the world. Scientists try to be objective, unsentimental, unemotional. They do not let their feelings get in the way of their observations of real things, facts, as they call them. They often work in laboratories or in other areas where they can carefully control what they are working on. They do not just wander out onto the dock at sunset and look at the world with wonder, as a poet might. Ideally, they are also both honest and humble. They always try to report their findings so others can check them out and then utilize them in their own work. They do not claim more than they can prove, and often even less. But they are very proud of their calling and prefer to talk to other scientists rather that anybody else, especially poets, who tend to make them feel uncomfortable, to put them down. (Of course poets also feel scientists return the favor.)

Second, science deals almost exclusively with things, not ideas or feelings; and with the external world and its workings, not inner states and their workings, despite the effort of some psychologists to be or seem scientific. The human body is considered to be a part of the external world; the soul is not. Therefore, scientists work to understand the body but not the soul. Most scientists doubt the soul exists. The solar system and the universe are also part of the external world, although we have little enough direct evidence of their mode of existence. Scientists tend to assume the basic conditions of nature on earth are the same everywhere in the cosmos.

Mankind is the only questionable part of external world in this sense. Scientists are generally reluctant to deal with the behavior of large groups of men and women. Thus economists, for example, struggle to be considered scientists, but usually in vain. The external world of scientists contains some things, like quanta, quarks, and quasars, that are fully as mysterious as angels and normally as invisible. But this does not trouble them, as they believe they can deal effectively with the elementary particles that they cannot see and according to the uncertainty principle never can see, but not with angles, which will probably never appear to scientists because scientists do not believe in them.

When you come right down to it, external world is anything that scientists can measure and describe in mathematical terms, and it excludes everything they cannot. This means the external world is a rather hazy notion, but the idea behind it is not hazy at all.

Third, science deals with whatever it deals with in a special way, employing special methods and a language for reporting results that is unique to it. The best known method, but not necessarily the most often employed, consists of experiment, which involves getting an idea—from where, most scientists do not question—framing it in a testable hypothesis;

and then testing the hypotheses in a controlled environment to find out whether or not it is valid. The environment must be carefully controlled so that extraneous elements do no intrude to invalidate the experiment, and so that others can repeat the experiment in the hope of arriving at the same result, which is the best evidence of its reliability.

But it is the language in which results are reported and in which the work itself is done and with which it is controlled—namely, mathematics—that is perhaps the most distinctive characteristic of all. Most scientists would say that if you cannot describe what you are doing in mathematical terms, you are not doing science, and they prefer to report their results in mathematical terms because doing so is much easier and quicker (for them) and because scientists all around the world can understand them.

It is also important that the work itself is done mathematically, which means that the observations being studied must be transformed into—or reduced to—numbers in the first instance, so they can be studied in a rational manner. The old idea of the earliest Greek scientists—that the world is essentially intelligible because it is somehow conformed to the human mind—is thus converted into the Pythagorean view that the world, at least the external world that is the subject matter of science, is essentially mathematical and thus intelligible because the human mind is essentially mathematical, too.

Wherever mankind has been able to measure things, which means to transform or reduce them to numbers, it has indeed made great progress both in understanding and in controlling them. Where human beings have failed to find a way to measure, they have been much less successful, which partly explains the relative failure of psychology, economics, and literary criticism to acquire the status of science.

Science was the major discovery, or invention, of the seventeenth century. Men of that time learned—and it was a very great, revolutionary discovery—how to measure, explain, and manipulate natural phenomena in the way that today we call scientific. Since the seventeenth century, science has progressed a great deal and has discovered many truths, and conferred many benefits, that the seventeenth century did not know. But it has not found a new way to discover natural truths.[1] For this reason, the seventeenth century is possibly the most important century in human history. It instituted irrevocable change in the way human beings live on earth. We can never go back to living the way we lived in the Renaissance, for instance. We can only wonder whether the change was in all ways for the better.

Aristotelian Science: Matter

In order to invent scientific method, thinkers of the seventeenth century first had to overthrow the world view of the greatest scientist who had lived up until that time, Aristotle. To understand what happened we have to know something about the world as Aristotle saw and described it. Two aspects of that world, in particular, concern us: matter and motion.

Every material thing, said Aristotle, has both a material and a formal aspect. Matter, in one sense, is a thing's potentiality. Matter in this sense does not exist by itself. In another sense of matter, it is the stuff out of which things are made. It is the wax that is shaped by the imposition of the form, to use an old image that was often employed by Aristotelians.

In out sublunary world, the world below the moon, beyond which things are considerably different, there are four kinds of stuff out of which things are made. Four elements, as the Aristotelians preferred to say. They are Earth, Water, Air, and Fire. I give them capital letters because none of them exists purely in our imperfect world, but always in mixtures that

are more or less earthy, more or less humid, more or less aerial, more or less fiery.

Heavy things are mostly, although never entirely, made of the Earth element. Lighter things have an admixture of Water, Air, or even Fire, which, like the other elements, joins with them in mixtures. Since the four elements never appear alone, in their essential purity, it is very hard to measure them. In a sense they are invisible. But it is obvious enough, the Aristotelians said, that a man has good amount of Earth in him, which makes him heavy, contributes to the strength of his bones, etc.; a good amount of Water, which produces his blood and other internal fluids; of Air, which he breathes in and out; and of Fire, which gives him his heat and is in a sense the essence of the life in him. And so with other material things beneath the moon.

Above the moon, that is, in the sun and the planets, the fixed stars and the great spheres on which they all move, there is a fifth element, a Quintessence, as it was called. The sun and the other celestial bodies are made out of the Quintessence, which exists in them in a pure state. The moon is mostly made out of the Quintessence, although there is a small admixture of the sublunary elements in it because of its proximity to the earth, which is mostly made of Earth. The proof of this is the markings on the moon, which are like the ravage made by time upon a beautiful face. It is important to remember that the quintessential element of which the celestial bodies are made is still matter. It is not what angels are made of, for example, because angels are nonmaterial, as is God.

Aristotelian Motion

The fundamental fact for Aristotle, the basic, underlying assumption of his physics, which was well and consistently structured, is that the natural state of all sublunary things, material and immaterial, is rest. Motion, as a consequence, is always either violent and

---

[1] It may not be strictly correct that we have not discovered any new ways to discover truths. See Chapter 13.

unnatural, or it is a natural correction of a previous state of imbalance, that is, a seeking, on the part of the body, for its place of rest. Once that place of rest is achieved, motion stops.

Earth, Water, and to a certain extent Air naturally seek a place that is downward, toward the center of the earth, which they would reach if they could, that is, if they were not stopped at some impermeable surface, like that of the earth itself. Fire seeks to fly upward to its natural place of rest, which is above us, but not infinitely far, that is, that place is well below the sphere of the moon. Air is often, Perhaps always, mixed with Fire, as well as with the heavier elements, and so its behavior is flighty and unpredictable. It goes up, it goes down, its movements being highly perturbed because of the odd mixture of elements within it. If Air were pure, it would rest in its natural place around us, with Water and Earth below it, Fire above, and there would be no wind.

Before discarding this picture of the would, consider how sensible it seems, and what a stroke of genius it was to arrive at it. In our experience, everything *is* at rest, unless it is seeking that natural place where it can find rest, as the river seeks the sea, the flame its place above us, or is forced to move by something else. When we force something to move—say, throw a ball—it soon rolls to a stop and will stay in the place it has found until we pick it up and throw it again. So it is with all material things lacking souls. We have no direct sensory experience of anything—anything at all—that does not seem to "desire" to find a place where it may rest.

And what of things that have souls, like animals and men? They, too, seem to seek a natural place, a home, ultimately a grave. For is not the grave the end and goal of all striving? The body seeks that goal. But the human soul strives for something else, atonement with God, the peace that God alone can give. That is the highest and strongest desire of the soul, even if sometimes, as Dante explains in the sixteenth canto of the *Purgatorio*, the soul wills not aright.

"My love is my weight," said St. Augustine, a statement that is unintelligible unless one understands Aristotle's universe, and then it is obvious. My body seeks the earth, because it is earthy. The element Earth predominates in it. But my spirit seeks a higher resting place. That is what it loves. The weight of my body draws me downward. The weight of my spirit is light, lighter than Air, lighter than Fire, and its lightness snatches it upward to its natural resting place, while my body rests in its long home.

In the sublunary world, then, there are rest and two kinds of motion: motion that is natural because it results from the "weight" of a thing, which always seeks its proper place ("proper" means "own"); and motion that is unnatural or violent, as Aristotle said, because it is the result of a force being applied to a thing. But what of the world above the moon? There is motion there, too! The sun and the planets move, the fixed stars circle the world once every twenty-four hours. What kind of motion is that?

This was a hard question, for beneath the moon all motion is in straight lines, unless some violent force turns a body out of the right path. Above the moon, the sun, the planets, and the fixed stars apparently move in circles. Are they forced to do so? We cannot assume that, said Aristotle and his Christian followers, for the heavenly bodies are perfect, and it would be imperfect to be pushed. Their circular motion must, somehow, be a natural motion.

The solution followed easily: the natural motion of the Quintessence is uniform circular motion, which differs from the motions of sublunary things as the heavenly bodies differ from those things. Immediately, all is explained. The heavenly bodies, or rather the spheres on which they move. Turn forever because that is their nature, and we see the result when we look up into the sky.

From time to time another theory was advanced, to the effect that angels drove the planets in their paths, effortlessly moving them forever in their appointed rounds. This theory, in fact, was widely accepted during the early Middle Ages. When Aristotle was rediscovered after 1000 it became clear how much better was his assumption of a natural quintessential motion that attached itself to a natural quintessential substance. The world made more sense that way. It was somehow more fitting, more beautiful, more perfect, and more the way God would obviously have made it. And so this theory that the planets moved in that way turned into dogma. To question this belief was to question God's design for the world.

The Revolt Against Aristotle

Galileo challenged Aristotle's theory of motion, thus producing the most famous moment in the history of science, but it was far from the first such event. The questioning had begun at least two centuries before Galileo was born.

Why did the questioning start? Aristotle's theory of motion explained the way things naturally fall and run downhill—a ball dropped from a tower, a river running to the sea—but it was much less successful at explaining what Aristotle called violent motion, This is the kind of motion that a body undergoes when it is thrown or hurled by some sort of machine like a catapult or a cannon. It was the invention and regular use of catapults, in fact, that may have led to such questioning. The traditional theory did not explain very well how they worked.

That may be difficult to understand, since we now have and entirely different theory of motion. But if you remember that Aristotle's law of inertia was based on the principle of rest, you will see the problem. Nothing moved, in his theory, unless it was pushed, or unless it was partaking of a natural motion, like the fall of an object toward the center of the earth, or the uniform circular motion of the heavenly bodies.

A projectile shot from a catapult was not moving naturally. While it was rising on the catapult's throwing end, it was obviously being pushed. But why did it

keep moving once it left the catapult? It was no longer being pushed. Why did it not drop straight down to the ground as soon as it was free to do so?

Aristotelians had answers to these questions, but they were inadequate, indeed, rather lame, The splendidly commonsensical theory of inertial rest broke down when it came to violent motion. For example, it was said that the air in front of the projectile became disturbed and rushed around and behind the projectile in order to fill up the vacuum caused by its passage, since "Nature abhors a vacuum." This frantic effort on the part of the air to avoid a vacuum pushed the projectile forward. And there were even more fanciful explanations.

Many thinkers gave it up as a bad job. Violent motion was just hard to explain, they said, but the theory in general was so obviously right that this should not matter a great deal. But some eminent theologians at the University of Paris were more skeptical. Since they were recognized authorities in theology, they could question with impunity a part of Aristotelian theory, knowing as they did how to save the remainder. This is what Galileo, later, did not want or know how to do.

Jean Buridan (1300–1358) was one of those Parisian theologians. Nicholas of Oresme (c.1325–1382) was another. They saw the problem clearly, and they came up with a solution. The catapult, they said, imparts a certain *impetus* to the projectile, which continues to move on its own until the impetus is spent.

Violent motion, in other words, is inherent; like natural motion, its principle is in the body that moves. Once the impetus has been imparted to the projectile by a violent force, the projectile no longer needs to be pushed. It keeps on going until (in the case of a cannonball or a projectile from a catapult) it falls to earth.

This was good as far as it went, but it did not go far enough. The problem of uniform circular motion remained, and the theologians did not see how to apply their insight to that problem. Also, to do so might be treading on dangerous ground.

There were several serious problems about the way the heavenly bodies moved, or were supposed to move. First, did the assumption of uniform circular motion save the phenomena, as the saying went? Did it explain what astronomers observed when they looked at the sky? For Ptolemy, the great Alexandrian of twelve hundred years before, uniform circular motion had been adequate to explain what he had been able to observe, and what his predecessors could hand down to him in the way of observation. But now the heavens had been watched with scrupulous care by a horde of astronomers over the centuries, Arabs and Greeks, Indians and Italians. When their observations were pooled and collated, it began to look as if the theory of uniform circular motions, even when the motions were combined in ingenious ways, would *not* save the phenomena.

The combining of uniform circular motions had been necessary for some time. The ancient Greek astronomers had been able to see, for example, that the apparent path of Venus in the heavens is not a uniform circle around the earth. The phenomena could be explained if one assumed that an ideal point circled the earth uniformly, which point was the ideal position of Venus, while the planet itself circled uniformly around that ideal point. This view accounted for the observed fact that Venus appeared to move forward in its orbit faster at some times than at others, and in fact sometimes appeared to move backward in its orbit, to retrogress. The uniform circular motion of Venus around its ideal point was called the epicycle of Venus.

As more accurate observation continued to be made by astronomers over the centuries, more epicycles were needed to explain the observations. Eventually, every planet needed an epicycle. Mars needed two, for only if it were assumed that the planet uniformly circled a point on an epicycle that in turn uniformly circled the ideal point of Mars could the perturbations of the observed orbit of the planet be

explained. Eve so, the theory of epicycles did not work perfectly, as the accuracy of observations continued to be improved. Besides, epicycles were not elegant. It was unpleasant to have to think of the heavens cranking around in such an unaesthetic manner.

But if the planets did not move in uniform circles around the earth, how then did they move? Was there any other kind of simple motion that would explain the appearances and could be called "natural"? There did not appear to be such a motion. At least no one could imagine it.

As time went on, there were many other problems that had not been solved. For example, why did the heavenly bodies move in the first place, whether in uniform circles or in some other way? The answer that had once been universally acceptable—that God wished them to move, and so they moved—had begun to be troublesome to the most adventurous minds. The assumption of the Quintessence was also difficult to accept. This was especially true of the quintessential motion itself. Many thinkers were beginning to be uncomfortable with a type of motion that is never observed on earth, where nothing ever moves naturally in a uniform circle. (On earth, if something moves in a circle it is because it is being *forced* to move in that way.) If angels or intelligences did not move the sun and the planets and the fixed stars—if they moved by themselves—then what was the cause of that motion?

In addition, there was the problem of the crystalline spheres on which the heavenly bodies were said to move. They could not move in empty space, because empty space, for several reasons—for example, that nature abhors a vacuum—was unthinkable. (Aristotle had quarreled with Democritus on this point.) These great spheres, which made heavenly, although inaudible, music as they turned, were invisible. That was all right. We certainly do not see them. But the epicycles, some on top of others, were also crystalline spheres, and it appeared as if some of the spheres had to intersect other spheres. But this was impossible, because the quintessential matter of which they were made was assumed to be impermeable, unchangeable, indestructible, and so forth.

Finally, there was a special problem about the fixed stars. They were supposed to move on a crystalline sphere outside the sphere of Saturn. (Beyond the fixed stars was the Empyrean, the abode of God.) Observations made since Ptolemy's time on stellar parallax had shown that this sphere, and all the stars on it, must be very far away. But if they were so far away, then the speed with which their sphere turned about the earth every twenty-four hours must be almost unimaginable great. In a sense, this was not a problem, as God could have arranged for it to turn as fast as he pleased. There was no limit on the divine power. Even so, the theory seemed difficult. And many men in several lands sought a simpler solution to the problem.

Copernicus

Nicolaus Copernicus was born in 1473 and lived most of his life in Poland. He received an excellent education in the universities of eastern Europe and by 1500 was already said to have mastered all the scientific knowledge of his time: medicine and law as well as mathematics and astronomy. He could have chosen any learned profession, but he selected astronomy.

The more he studied and thought about the reigning Ptolemaic-Aristotelian theory of the heavens, the more it troubled him. The theory seemed complicated. Was it unnecessarily so? For example, if the earth rotated, that would explain why the fixed stars revolved around the earth every day, and the problem of their rapid motion would be solved. They would not have to move at all. And if the earth revolved around the sun, instead of the sun around the earth, that would simplify the problem of explaining the planetary orbits.

Copernicus studied all the old Greek astronomical texts he could find. He discovered that a rotating earth

and a heliocentric system had been proposed by more than one ancient Greek astronomer. Was it possible to make a small change in the assumptions, and obtain a major improvement? Copernicus began to think so.

He was timid, however, and he did not publish the book he was writing, *On the Revolution of the Heavenly Orbs.* He delayed and delayed. In fact, he only permitted the book to go to the printer when he was on his deathbed. A copy of his great work was brought to him on the day he died in 1543.

He had been afraid of religious controversy and of what the orthodox Aristotelians would say about his ideas. In fact, they said surprisingly little, partly because an introduction to his book, written by a friend, emphasized that the theory was only a hypothesis, designed to simplify certain mathematical difficulties. Copernicus was not actually saying that the earth *did* rotate once a day and *did* revolve around the sun once a year, the introduction asserted, although careful readers of the book realized that Copernicus actually *was* saying that. And so the new theory did not produce the intellectual revolution that Copernicus may even have wished for, although he was afraid to bring it about during his lifetime.

Perhaps the main reason Copernicus did not carry out the so-called Copernican revolution is that he had been careful to retain two important features of the Aristotelian system. One involved uniform circular motion. The other was quintessential matter, for which such motion was said to be natural. Theologians, therefore, as well as some astronomers, could believe that nothing really important had changed.

Tycho Brahe

This great Danish astronomer knew much had changed. Born in 1546, Tycho was abducted by his childless, wealthy uncle at an early age; after the initial family shock had been overcome, the uncle raised the boy, saw that he received an excellent education, and made him his heir. Tycho disappointed his benefactor in one respect. Despite his uncle's wish that he become a lawyer, he instead insisted on a career in astronomy. Inheriting the estates of both his father and uncle before he was twenty-five, he became independently wealthy and able to do what he wished with his life.

Aided by further financial assistance from the king of Denmark, Tycho established his own observatory on an island near Copenhagen, where he set about doing what he considered his life's work, namely, to correct all of the existing astronomical record, which he knew were grossly inaccurate. Perhaps the most dramatic event of his life was discovery, in 1572, of a nova in the constellation of Cassiopeia. He observed the bright new star over a period of months and in 1573 published a monograph on it that made him instantly famous and instantly controversial.

New stars were not supposed to come into being in the Aristotelian and the Christian universe. The world below the moon was chaotic, imperfect, and unpredictable changeable. That was an acceptable although not a very desirable situation. Basically, it was the fault of the Devil, who had disturbed God's originally perfect world by tempting Eve and Adam into sin. Above the moon, however, the heavens did not change. They continued to reflect God's immutable love for the world and mankind. The theologians, therefore, after duly investigating Tycho's monograph, concluded that the paper and its author were in error. The new star was not really new. It simply had not been observed before.

Tycho was not surprised, nor was he terribly disappointed. He was personally wealthy, and Denmark was a Lutheran country. His king was a staunch Protestant and cared little more than did Tycho for the criticisms of Roman Catholic divines. In any case, Tycho continued to desire more than anything else to leave to posterity a collection of astronomical observations sufficiently accurate so that future generations would be able to depend on them.

After 1588 a new king provided Tycho less financial support, and he finally had to give up his beloved observatory and settle in Prague, where in much reduced circumstance he was able to complete his work with the assistance of a young student, Johannes Kepler, to whom, at his death in 1601, he left all of his astronomical data. What Kepler did with them we will learn in a moment.

Gilbert

William Gilbert, an Englishman, added a crucial piece of information to the growing body of knowledge that would eventually overthrow the fixed and unchanging Aristotelian world picture and replace it with another. Like his contemporary William Harvey (1578–1657), the discoverer of the way the heart works to pump blood through the arteries and veins of the body, Gilbert (1544–1603) was trained as a physician and practiced medicine with much success. But it was his scientific hobby that made him famous. He was fascinated by lodestone, the mineral now called magnetite that possesses natural magnetism and is found in many places throughout the world.

Gilbert studied lodestones of all kinds, shapes, and powers of magnetism. His most important discovery was that the earth itself is a magnet, which he deduced when he observed that a compass needle dips downward when it finds the magnetic north (in the northern hemisphere). Gilbert also suspected that the earth's gravity and its magnetism were connected in some way, but he never understood how.

England, like Denmark, was Protestant, and Gilbert was supported by another Protestant monarch, Queen Elizabeth I. He therefore was able to proclaim to the world his remarkable modern ideas. He argued forcefully for Copernicus's heliocentric picture of the solar system and concluded that not all of the fixed stars were the same distance away. But his most provocative idea suggested that the planets must be held in their orbits by some kind of magnetism. No one else understood the implications of this suggestion at the time; nor, in fact, did Gilbert himself understand very well what he was proposing.

Kepler

Johannes Kepler was born in Württemberg in 1571 and died in 1630. Although the son of poor (although noble) parents, he received and excellent and wide-ranging education in Lutheran schools and at the University of Tübingen. He hoped to follow a career in the church, but he wrote a paper on an astronomical subject that came to the attention of Tycho Brahe, now at Prague, and Tycho invited the young man to join him as his assistant. After much soul-searching, Kepler accepted, and when Tycho died the next year, in 1601, Kepler was appointed imperial mathematician in his place and inherited Tycho's large body of accurate astronomical observations.

Kepler evidently felt that he had inherited more than just data. He also began to view more positively Tycho's unorthodox views, some of which Kepler now recognized for the first time. Tycho had published papers disputing the theory of the crystalline spheres on which the planets were supposed to move. Kepler followed up his argument that the planets moved freely in space and incorporated it in his own works. Like Tycho, Kepler also came to view Copernicus's heliocentric theory as more than a mere hypothesis, and he published papers arguing that no description of the world with the earth instead of the sun at the center could be accepted. But his greatest contribution was a set of three laws of planetary motion that solved the problem of epicycles and eccentric orbits once and for all. The three laws are still valid and are called by his name.

The first of the new laws made a substantial change in the Aristotelian system, for it asserted that planetary motion is not uniformly circular. The planets do not travel in eccentric circles around the sun, but in ellipses, with the sun at one of the two foci of the ellipse. Kepler's ellipses were very close to circles, which explained why the previous assumption of circular orbits had adequately explained the phenomena as long as observations remained relatively inaccurate. The new assumption was correct within the limits of observational accuracy of the time and required no further adjustments, no eccentricities, no epicycles, no tricks of any kind.

Kepler's second law of planetary motion asserted that a radius vector joining a planet to the sun sweeps out equal areas in equal times. What this means is that in a certain time, a planet will travel more quickly along its orbit when it is closer to the sun than when it is farther away from it. This brilliant insight, a major inspiration to Newton, applies to all bodies moving in fields of force, not just planets. It explained most of the discrepancies between astronomical theory and observation. Unfortunately, the idea remained an intuition in Kepler's mind. He knew it was correct, and it is, but he did not really understand why.

The third law asserted a mathematical relation between the periods of revolution of the planets and their distance from the sun. Discovering this law was a remarkable achievement considering the primitive instruments Kepler had at his disposal.

Kepler spent many years not only advancing his ideas about these laws and preparing Tycho's tables of observations for publication, but also mulling over what he recognized as the great remaining unsolved problem of planetary motion: the motivation whereby the planets revolve around the sun. What holds the planets in their orbits, and what drives them ever forward?

He realized that the speculations of Gilbert about the earth as a magnet must have something to do with the answer to the question, but he never understood what it was. He discarded almost all of the Aristotelian celestial baggage, including the idea of intelligences that guided the planets in their eternal round. He was also able to accept the idea of a force acting at a distance upon the planets, with no physical entity between the sun and the planets that it controlled. But he could not discard one crucial Aristotelian assumption, that of inertial rest. He came so very close to discovering the secret that made Newton the premier scientist, but he missed it because he thought the planets would stop moving unless something kept pushing them, and he could not imagine anything doing that other than Gilbert's magnetic force, he was very slightly wrong on both counts, and so he is remembered as an important precursor to Newton, but no more.

Galileo

Galileo Galilei was born in Pisa in 1564 and died in Acetri, near Florence, in 1646, He was a Roman Catholic and he lived in a Catholic country. That was one major difference between him and Tycho, Gilbert, and Kepler.

He studied at Pisa and taught mathematics at Padua. He was the leading mathematical physicist of his age, not just because he was very good at geometry. He was also the first modern man to understand that mathematics can truly describe the physical world. "The Book of Nature," as he said, "is written in mathematics."

As a young man Galileo conducted elegant experiments showing the inadequacy of Aristotle's theory of violent motion. He accepted Buridan's impetus theory and proved that projectiles shot from guns follow parabolic paths as they fall to earth. He studied the pendulum and showed that it, like the planets, sweeps out equal areas in equal times. All of this was theoretical work, and it did not get him into trouble. His troubles began in Venice in the spring of 1609, when he learned of the recent invention of the telescope. Upon returning to Padua he made a telescope of his own and quickly improved it to the point where it was better than any existing instrument. During the

summer and fall of 1609 and the winter of 1610, he undertook a series of observations.

The first thing Galileo looked at with his telescope was the moon. To his great wonder he discovered that the surface of the moon was not smooth. There were mountains and valleys corresponding to the features that had always been seen but never before understood. This was not so shocking, as it had always been supposed that the moon was not made entirely of quintessential matter. He looked at Jupiter, and discovered its moons. Jupiter, then, was a little solar system which in turn revolved around a larger body. Finally, he turned his telescope on the sun and discovered curious spots on the sun's surface. These dark areas were not constant. He could see them change shape and position from night to night, from month to moth.

The heavens, therefore, were not immutable and indestructible. Mountains and valleys had been formed on the moon by processes that, Galileo concluded, must be similar to those that operate on the earth. Jupiter was a miniplanetary system, and there might be many more such systems that he could not see as yet with his primitive instrument. And the sun was a living thing that was subject to change and did so before his eyes.

In 1611 Galileo went to Rome to describe what he had seen to the pontifical court. He took his telescope with him. Many were impressed by his finding, the meaning of which they did not at first comprehend. But he demanded that they open their eyes to those consequences. Among other things, he said he could prove mathematically that the earth went around the sun and not the sun around the earth, that Ptolemy was wrong and Copernicus right. And, he insisted, his telescopic observations proved that the heavens were not basically different from the sublunary world. There was no such thing as the Quintessence. All matter, everywhere, must be the same, or at least very similar.

You can prove no such thing with your mathematics, said Cardinal Robert Bellarmine (1542–1621), chief theologian of the Roman Church. He reminded Galileo of the time-honored belief that mathematical hypotheses had nothing to do with physical reality. (It was this belief, held by the Church for centuries, that had protected Copernicus's work from oblivion.) Physical reality, the cardinal said, is explained not by mathematics but by the Scriptures and the Church Fathers.

Look through my telescope and see for yourself, said Galileo. Bellarmine looked, but he did not see.

Why were Cardinal Bellarmine and the Dominican preachers whose aid he enlisted in a campaign against Galileo unable to see what Galileo saw, and what we would see if we looked through that telescope? Their eyes were physically the same as ours, but they did not see as ours would.

They deeply believed in the Ptolemaic system and the Aristotelian world order. But not because they were physicists who thought that those theories better explained the phenomena. They knew little or nothing about the phenomena. They believed in the old theories because the theories supported even more deeply held beliefs. And to question those deepest beliefs was to bring their world crashing down around their heads. They could not face that possibility.

St. Augustine, more than a thousand years before, had described in *The City of God* the distinction between the two cities, the heavenly and the earthly, which could be said to define the life of man and the pilgrimage of his spirit. Augustine's distinction, certainly, had been allegorical only, that is, he had not thought that one could actually see, except with the mind's eye, either the City of Man or the City of God.

But over the centuries those great images had taken on a kind of reality that proved more powerful than what one might see before one's very eyes. The City of Man was here, beneath the moon, It was earthy, material, strong-tasting and strong-smelling. It was the ordinary life of man. But in the heavens,

at night, the City of God became visible to those who had eyes to see it. It shone there, unchangeable, indestructible, always beautiful. It was the promise of God to the faithful, the ark of the Christian, not the Jewish, Covenant.

It was the loveliest, the most desirable thing in the universe. To call it in question, to destroy it, to bring it tumbling down, was unthinkable. Anyone who threatened to do so had to be stopped, and if necessary burned at the stake. Even if he should be the world's greatest scientist.

Galileo had little or no interest in the City of God of St. Augustine. He was a good Christian, but his faith was as simple as his mathematics was subtle and complex. He went to church, he took communion, and during the sermon he did computations in his head. He watched the hanging lanterns of the cathedral swinging lazily in the breeze and worked out theories about the pendulum. For him, too, the heavens possessed an extraordinary splendor, but it was very different from the splendor of Cardinal Bellarmine's divine city. The heavens held a promise for him, too, but the promise was different. They could be studied, understood, even controlled in some way. So Galileo dreamed.

Bellarmine was much at fault for not trying to understand Galileo, for not recognizing the kind of new man he was, who would never willingly harm the Church, a good Catholic who would not allow himself to be wooed by the Protestants, as Bellarmine feared. Another time-honored doctrine supported Galileo, to wit, when the Scriptures conflicted with scientific truth, the Scriptures had to be interpreted allegorically, to avoid "the terrible detriment for souls if people found themselves convinced by proof of something that it was made then a sin to believe. "This sophisticated argument had probably been suggested to Galileo by one of his theologian friends. He would not have thought of it by himself. But Bellarmine ignored the argument, although it would have given him a good fallback position. He forged ahead, heedless of the political consequences of prosecuting and condemning Galileo, perhaps even to death.

Galileo was also much at fault for not trying to understand Bellarmine and those who thought like him. The dispute was not merely scientific, and it was certainly not about a particular scientific truth, such as whether the sun goes around the earth or the earth around the sun.

It was about science itself, about the role it ought to play in human life, and particularly about whether scientists should be permitted to speculate with absolute freedom about reality. Even more than that, it was about the City of God, which could never again be viewed in the same way if Galileo was right.

Or rather, if he were allowed to say he was right in the way he wanted to say it. Everyone knew he was right in a way; his hypotheses were much more satisfactory than anyone else's. But Galileo wanted to go beyond mere hypotheses. He insisted that what he could prove mathematically and by means of his observations was *true*, and that it could not be questioned by anyone except a better mathematician or a better observer.

The Church had no authority, he was saying, to describe physical reality. But then what authority would remain to the Church? If the Church could no longer say, in every sphere and not just that of the spirit, what is and what is not, would not the Church be reduced to a mere adviser of souls? And if that were to happen, the danger existed that millions of souls would cease to ask for the Church's advice. And would not most of them then, in all likelihood, go to hell?

So Cardinal Bellarmine argued. His understanding of the choice that mankind faced was clear. Galileo was condemned to be silent, and for the most part he was, Bellarmine became a saint. He was canonized in 1930. But in the long run, of course, Galileo won. The Church has been reduced to an adviser of souls, in the Western world at least, and science has been elevated to the position of supreme authority.

Bellarmine failed because he was not a good enough theologian. He should have read Augustine better and seen that the two cities are only allegorical. They are not real in the same way as what one sees through a telescope. St. Augustine, and many who understand him better, had always been able to juggle two kinds of reality, which might be said to correspond to the two cities. Let Galileo be the authority in the City of Man. The Church could remain the authority in the City of God. Because the Church wanted both kinds of authority, it ended up with neither.

Now, when we look up at the stars on a clear, dark night, we see a splendid vision, but it is not the vision that mankind once saw there. We have both gained and lost because of that.

Descartes

René Descartes was born in La Haye, France (now called La Haye-Descartes) in 1596, and died in Sweden in 1650 from a severe cold brought on by the requirement that he conduct philosophy lessons at five o'clock in the morning during northern winters. He had always preferred to lie in bed, and, besides, he hated the cold, but his patron, Queen Cristina, insisted on philosophy at five, and he could not say her nay. Such ironies make the history of science and amusing subject of study.

Other ironies illuminate the biography of René Descartes. He possessed a deep Catholic faith, but his writings did more to undermine the authority of the Church than the words of any other person. He created a scientific methodology that would revolutionize not only science but also the way mankind lives in the world. But his own views of things were often wrong, and in some cases so disastrously conceived that they impeded French scientific progress for two centuries, since French thinkers tended to believe that they must follow Descartes, whether they understood him or not. Similarly, English insistence that Newton's terminology for the calculus was better than Leibnitz's—which was nonsense, despite the fact that Newton had certainly been the first to invent the calculus—set back English mathematics for more than a century. Most ironic of all, Descartes's search for certainty was based upon the principle that everything should be doubted. This was an odd idea, but in fact it worked.

Descartes received the finest Jesuit education that could be obtained in the Europe of his time, an education that included an exhaustive study of Aristotelian logic and physical science. But when he graduated, at the age of twenty, he was in despair, for he felt that he knew nothing with the certainty with which he desired to know everything. Or rather, he knew nothing with that certainty except some mathematical truths.

In mathematics, he felt, it was possible to know things, for you started from axioms that possessed the character of indubitable certainty, and built from there, by small steps, a structure that possessed the same character. Such certainty adhered to nothing else, he thought, not to any other science, not to history, not to philosophy, not even to theology, despite the claim of the last to the highest certainty available to man's mind.

By 1639, after wide travels, much reading, and a voluminous correspondence with the most progressive thinkers of Europe, Descartes was ready to write a kind of summa of his philosophy which would organize all knowledge into one great structure, based on a universal method that led to certainty. But in that year he learned of Galileo's condemnation, and he decided he had better not write *that* book. Instead, he wrote *The Discourse on Method*, which concentrated on the method only, and left to others the work of applying it to discover controversial new truths. Nevertheless, even the Discourse got *Descartes* into serious trouble.

It is an absolutely astonishing book. In it, in French that exemplified the clarity and distinctness of the author's thought, he recounted the history of his intellectual development, how he began to doubt whether what he had been taught was true, and continued to doubt until he arrived at the simple conclusion that all might be doubted except one thing, namely, that he, the doubter, existed *because* he doubted. (*Dubito ergo sum.* "I doubt; therefore I am.") He then proceeded to discover a method of achieving similar certainty in other realms, based on the reduction of all problems to a mathematical form and solution. Thereupon he proved the existence of God mathematically and at the same time showed how God had created a world that would run forever without his assistance, like a huge, complex, and ornate clock. And he managed to do all of this in twenty-five pages. An amazing performance.

The method itself was the important thing. To understand some phenomenon or set of phenomena, first rid your mind of all preconceptions. This is not easy, and Descartes was not always successful in doing it. Second, reduce the problem to mathematical form, and then employ the minimum number of axioms, or self-evident propositions, to shape it. Then, using analytic geometry, which Descartes invented for the purpose, further reduce the description of the phenomena to a set of numbers. Finally, applying the rules of algebra, solve the equations that result, and you will have the certain knowledge that you seek.

Galileo had said that the Book of Nature is written in mathematical characters. Descartes showed that these mathematical characters are simply numbers, for to every real point there can be attached a set of Cartesian coordinates, as Leibinitz was to name them, and to every line, whether curved or straight, and to every body, whether simple or complex, corresponds a mathematical equation.

Human beings are not mathematical equations, admitted Descartes, but it is sufficient for many proposed to describe them as such. In the case of the machines that we call animals—they are machines, he said, because they lack souls—the equations are sufficient for any and every purpose. For all other machines, including that greatest of machines, the universe, the equations are certainly adequate. It only remains to solve them. That may be very difficult, but by definition it is possible.

The Cartesian worldview affected everyone, not least those who hated and condemned Descartes for it. Pascal could not forgive him for not needing God except to start the universe going, and the Catholic theologians, by now as desperate as Descartes had been on his graduation day, felt in necessary to condemn him for a dozen kinds of heresy and to place his Discourse on the Index of forbidden books. But even they coveted the certainty that Descartes and his method promised. If only theology could be reduced to geometrical form!

That cannot be, despite the effort of Spinoza to make it so, for theology deals with an immaterial world that mathematics cannot enter. This is the main characteristic of theology that had attracted the passionate interest of the best thinkers for a thousand years. Now, suddenly, it ceased to be attractive. The world of the immaterial, which had been supremely interesting, suddenly ceased to be interesting at all. It is one of the most radical changes in the history of thought.

There were major consequences. Descartes's triumph consisted of his invention of a method for effectively dealing with the material world. His disastrous failure came about because his method could effectively deal only with the material world. Thus, living as we do in the wake of his great invention, we inhabit a world that is resolutely material, and therefore in many respects a desert of the spirit.

Before Descartes, theology had been the queen of the sciences, mathematical physics a poor relation. After him, the hierarchy was practically reversed. Not for an instant had there been a balanced universe of knowledge. Is such a thing possible? That is an important question for the future to decide.

Newton

In addition to everything else, Descartes made Newton possible. Isaac Newton, the preeminent scientific genius of all time, was born in Woolsthorpe, Lincolnshire, England, on Christmas Day of 1642. He studied at Cambridge and, upon graduation, was offered the post of professor of mathematics. Isaac Barrow, his predecessor, who had been his teacher, resigned to make way for his extraordinary pupil.

Before graduating, Newton had discovered (that is, he stated it without proving it) the binomial theorem. That would have made the career of most other mathematicians. In was only the beginning for him. In 1666, when he was twenty-two, the plague that had decimated London attacked Cambridge, and he retired to his farm in the country. Farming did not interest him, and he equipped a room with instruments for experiments on light. Forty years later the revolutionary results that he discovered would be described in his *Opticks*. But this year held even more revolutionary thoughts for Newton.

All intellectual roads led to that room in Lincolnshire. Gilbert had performed his experiments on the lodestone and had hypothesized an earth exerting an attractive force, like a magnet. Galileo had not only seen the moons of Jupiter but had also studied falling objects and had accurately measured the force of gravity at sea level. Descartes had shown how to apply mathematical methods to physical problems. Kepler had described the elliptical paths of the planets, and had assumed a strange force, emanating from the sun, that drove them in their courses. And the Parisian theologians had proposed the impetus theory of violent motion, which called in question Aristotle's assumption of inertial rest. Looking back, it does not seem to have been difficult, what Newton did. One might think almost anyone could have done it, having all those pieces before him.

To say that is not to detract from the genius of Newton. For although all the pieces of the puzzle lay before him, so that he only had to put them together, it remains true that what was required was a mind entirely free of traditional prejudices and capable of seeing the universe in a new way. There have been few such minds, and in science, very few.

More was required than pushing around pieces of a puzzle. First, Newton had to be very well educated in the science of his time. Then, he had to be an accomplished experimenter and handler of instruments. Finally, like Descartes, he had to be an exceptional mathematician, capable of inventing the new mathematics needed to solve the problems he set for himself. Descartes's analytic geometry had been effective in dealing with a static universe. But the real world was constantly in motion. Newton invented the differential and integral calculus to deal with that phenomenon. Perhaps no other single gift to science has ever been more valued.

Gilbert plus Galileo plus Kepler plus Descartes add up to Newtonian mechanics. A new set of laws of motion was the first stage of the process. They are stated with consummate simplicity at the beginning of Newton's great book, *Mathematical Principles of Natural Philosophy (Newton's Principia,* for short). They define a universe utterly different from Aristotle's.

The first law asserts that every physical body continues in its state of rest, or of uniform motion in a straight line, unless it is compelled to change that state by a force of forces impressed upon it. A moving projectile continues to move in a straight line unless it is retarded by the resistance of the air or its path is curved downward by the force of gravity. A top, set spinning, continues to spin unless it is retarded by friction with the surface on which its point spins, or by the resistance of the air. The great bodies of the planets and comets, meeting with less resistance or perhaps no resistance in empty space, continue their motions, whether straight or curved, for a much longer time.

This law obliterated the Aristotelian concept of inertia. There is no such thing as the "natural state of rest" of a body. If a body is at rest, it will remain at rest

forever unless it is moved. If a body is moving, it will continue to move forever unless it is stopped, or its movement is changed in speed or direction by some force impressed upon it. Thus, no motion is "natural" and opposed to some other kind of motion that is "violent". Nor does one kind of motion have to be explained differently from other kinds. It follows, of course, that there is no such thing as quintessential motion, "naturally uniform and circular." Uniform motion in a circle is possible, but it is no more nor less natural than any other motion. Like all motions, furthermore, it is explained in terms of the inertia of bodies and the forces impressed upon them.

Newton's second law of motion asserts that a change of motion is proportional to the force impressed upon the body and is made in the direction of the straight line in which the force is impressed. A greater force induces a greater change of motion, and multiple forces produce a change that is a combination of the different strengths and directions of the forces. Analysis of the composition of forces is always possible using ordinary Euclidean geometry.

Ordinary Euclidean geometry cannot explain how the continuous impression of a force upon a body moving in a straight line can make the body follow a curved path, for example, a circle or an ellipse. The example was of the first importance, for all orbits are curved in the solar system. Newton made the assumption that a curved orbit could be conceived mathematically as made up of an indefinitely large number of indefinitely short straight line, joined to one another in a string around the center (or focus) of the orbit. In mathematical terms, the curved orbit could be considered the "limit" of a process of reduction or differentiation, in which the individual segments became each as small and as close to being mere points as desired, and of integration, in which the totality of all the segments came as close to being the smooth curve of the orbit as wished. That is the method of the calculus so far as it can be described in words and not mathematical symbols.

The third law of motion asserts that to every action there is always opposed an equal reaction. Or, the mutual actions of two bodies upon one another are always equal although directed in opposite directions. "If you press a stone with your finger," Newton says, "the finger is also pressed by the stone." And, by this third law, if you blast heated air out of the rear of a jet engine, the airplane to which the engine is attached will move forward in the opposite direction. Further, if one body revolves around a second each other. The velocities need not be equal; if one body is much bigger than the other, it will move very slowly, while the other moves relatively very quickly. But the total motions will be equal.

Curiously, this gave the final solution of the ancient puzzle: does the sun go around the earth, or the earth around the sun? They go around each other, and Ptolemy and Copernicus were both right, though for the wrong reasons.

Taking the three laws as given, then, let us suppose the planets in motion. They will remain in motion unless they are hindered by some force. The force need not stop them altogether. This force may only deflect them out of the straight line of their inertial paths. It may, indeed, deflect them into elliptical paths. By the traditional geometry of conic sections (going all the way back to Apollonius of Perga, in the third century bc; nothing new here) it *will* deflect them into elliptical paths (let us call them orbits henceforth) if the force is centripetal—that is, if the force attracts the planets *inward*, away from their inertial tendency to fly away from the center in straight lines—and if this centripetal force varies as the inverse of the square of the distance between the planets and the body exerting a force upon them.

Suppose that body to be the sun. What might that centripetal force be? Gilbert and Kepler had speculated that it must have something to do with the natural magnetism of the earth, but they were not in possession of Galileo's measurements of the force of gravity at sea level. Factor in those numbers, and the mysterious force is discovered. It is no other than gravitation, the force that holds the moon captive in its course around the earth, and allows the moon to control the ocean tides, that drives the solar system in its stately rounds, and that makes ripe apples fall to the ground or upon the head of an unsuspecting mathematician lying beneath the tree.

Newton claimed that he had understood all this while he was spending his enforced vacation in Lincolnshire in 1666. It seemed so simple to him, he said, that he told no one about it for twenty years. In the meantime he did other work that interested him more. When his *Principia* finally appeared in 1686, it made the world gasp. The greatest problem in the history of science up to that, the problem of how and why the universe worked as it did, had been solved. The poet Alexander Pope wrote:

> Nature and Nature's Laws lay hid in Night;
> God said, Let Newton be: and all was Light.

Rules of Reason

Isaac Newton was by nature a humble man, although a crusty one who often got into battles with his scientific colleagues. He once said to a biographer, "I do not know what I may appear to the world, but to myself I seem to have been only a boy playing on the sea-shore, and diverting myself in now and then finding a smoother pebble or a prettier shell than ordinary, whilst the great ocean of truth lay all undiscovered before me."

The image is as famous as it is intriguing. And probably it is even more accurate than Newton knew. That is, he was correct in admitting that he did not know a great deal compared to what could be known, even if he knew more than any other man of his time. And he was also correct in judging himself comfortable in his ignorance. The great ocean of truth lay all before him, but he did not even wish to stick his toe into it, to say nothing of shoving off from the shore with the goal of reaching the other side.

Book Three of Newton's *Principia* bears the awesome title, "The System of the World." It opens with two pages headed "Rules of Reasoning in Philosophy." We are to understand, first, that by "philosophy" Newton means "science." We may also understand that here is Newton's response to Descartes, his great footnote, as it were, to the *Discourse on Method.*

What are these rules of reasoning in science? There are only four. The first is this: We are to admit no more cause of natural things than such as are both true and sufficient to explain the appearances. This is a restatement of the logical principle first enunciated by William of Ockham in the fourteenth century, and now known as Ockham's Razor: "What can be done with fewer is done in vain with more." Newton, waxing a bit poetical, explains it thus:

To this purpose the philosophers say that Nature does nothing in vain, and more is in vain when less will serve; for Nature is pleased with simplicity, and affects not the pomp of superfluous causes.

The second rule asserts: Therefore to the same natural effects we must, as far as possible, assign the same cause. "As to respiration in a man and in a beast," Newton adds; "the descent of stones in Europe and in America; the light of our culinary fire and of the Sun; the reflection of light in the Earth, and in the planets."

Rule three answers a query that had plagued Aristotelians for centuries. It asserts, the qualities of bodies which are found to belong to all bodies within the reach of our experiments, are to be esteemed the universal qualities of all bodies whatsoever. As an example, Newton says, if the force of gravitation may be found to operate within the solar system, as it seems that it does, then we can—in fact we must—"universally allow that all bodies whatsoever are endowed with a principle of mutual gravitation."

The fourth rule of reasoning is , in Newton's view, perhaps the most important of all. The entire rule should be quoted:

In experimental philosophy [that is, science] we are to look upon propositions inferred by general induction from phenomena as accurately or very nearly true, notwithstanding any contrary hypotheses that may be imagined, till such time as other phenomena occur, by which they may either be made more accurate, or liable to exceptions.

Newton writes, "This rule we must follow that the argument of induction may not be evaded by hypotheses."

Newton loathed hypotheses. He saw in them all the egregious and harmful errors of the past. By "hypotheses" he meant the kind of explanations that the Scholastics had dreamed up to explain natural phenomena, the theory of the Elements, the assumption of the Quintessence, and the tortured explanations of so-called violent motion, which even the Parisian theologians had not been able to accept. And he was more than willing to admit what he did not know.

The most important things he did not know was the cause or causes of gravitation. That the earth and the other planets were held in their courses by the sun's gravity he had no doubt, but he did not know why. But "I frame no hypotheses," he declared; "for whatever is not deduced from the phenomena is to be called an hypothesis," and hypotheses "have no place" in science.

The four rules of reasoning, and the added prohibition against hypothesizing, that is, offering explanations not directly supported by experiments, could be said to define the scientific method as it has been practiced since Newton's time and as it is still practiced, for the most part, today.* Newton's rules established a new paradigm, to use a term employed by eminent historian of science, Thomas S. Kuhn, in *The Structure of Scientific Revolutions* (1962). The new paradigm inaugurated the age of science. The most valuable and useful tool for acquiring knowledge ever invented had been distributed among men, and with it they would proceed to try to understand everything they could see and many things they could not, as well as control the world around them in heretofore unimaginable ways.

Newton, with all his brilliance, did not understand why the force of gravity acts as it does; that is, he did not know what gravity is. Nor do we. He only knew that it acted the way it did. He was right about that, to his eternal credit. But the reasons of things, as Pascal might have called them, still lie hid in night.

That is partly the fault of Descartes, who made the search for them perhaps permanently unpopular. Partly it is the fault of Newton himself. His astonishing, brilliant success blinded the world to all those many things that it still did not know, and might never know. It is mostly the fault of the world itself, which is a harder thing to understand than mankind would like to believe.

The Galilean-Cartesian Revolution

Before moving on to the age of political revolutions, a word should be said about the names that are given to revolutions of all kinds. Often, the wrong person receives the credit or the blame. We shall see more examples of this in the next chapter. But a notable instance can be found in this chapter.

It has become customary to refer to the revolution that occurred in the seventeenth century—the revolution in ways of knowing that led to the establishment of science as the ultimate authority about material reality—as the Copernican Revolution. But this, I think, is unjust.

Copernicus, if in fact he desired to bring about a major change in thinking about the world, was afraid to produce it in his lifetime. He may never have had any such idea. Furthermore, his proposal that the earth revolves around the sun instead of the sun around the earth was not a revolutionary idea at all. Half a dozen ancient Greeks had said the same thing.

Other men had considered the idea. In itself, it was not a major change.

We say that it was, invoking the supposedly important notion that man was the center of universe before Copernicus, and not so thereafter. But this is far from the truth. As we have seen, man became the center of the universe, in any meaningful sense, with the Renaissance (with the discovery of perspective in painting, for instance), and they did not cease to be so at the end of the seventeenth century, when Newton's *Principia* appeared. That book, in fact, only solidified man's central position, as did all the scientific progress that followed it.

Today, when we look up at the night sky and know how many billions of stars and galaxies there are, and how tiny is our sun and its even tinier system of planets, of which the earth is far from the largest, it may not make us feel small or insignificant. Instead, it may make us feel strong and good, because we understand all that. Science exalts us; it does not belittle us.

Galileo was a very different man from Copernicus. For one thing, he was not afraid of the controversy that he knew his new ideas would produce. He was also not at all ignorant of the true meaning of what he was saying. He intended to replace the authority of the Church with another authority, because he believed the new authority—that of science—to be preferable in many ways. He did not fudge, as Copernicus had done. He really wanted to bring about a revolutionary change in the way men thought about things.

So did Descartes. He shared many of the mental characteristics of Galileo, although he was personally not so courageous. He was also more arrogant, which makes him not so likable. But he, too, knew what he was doing, as Galileo did and Copernicus did not.

If the revolution of the seventeenth century must be given the name of a man, then it ought to be called the Galilean Revolution, or, perhaps even better, the Galilean-Cartesian Revolution. Newton's name should not be used. He did not see himself as causing any very great change in thought. He was merely carrying forward the work of great men before him, and if he seemed the greatest of all, as indeed he was, he was not essentially different from them.

Unfortunately, the term "Galilean-Cartesian Revolution" does not sit very well on the tongue. And such things are important. Copernican Revolution sounds a good deal better. And so that is the name that historians will continue to use. But when I see it, I remember that Galileo and Descartes deserve much more credit than Copernicus.

The Game of Science

By Garvin McCain and Erwin M. Segal

Scientific Methodology

We have seen that science has as its object of study the natural world and that scientists strive to explain the facts of nature. We have discussed briefly what facts and data are and what constitutes an explanation. Our next problem is to examine how the scientist goes about finding facts and establishing explanations of them.

When scientists are at work they are not likely to gather data by casually observing the world around them. They may start by casual observation, but data obtained in this manner rarely show the relationships necessary for a powerful explanation. Such an explanation requires precise and reliable observations. Casual observation is likely to be contaminated by many extraneous factors.

Obviously, the explanation of planetary motion established by Kepler could not have been done without carefully controlled observation. The information he used had to be detailed and accurate. In addition to all his hard work, he had very good luck. In planetary motion, the effects of all variables except those caused by the sun are quite small. The elliptical orbit, which describes the path that a planet would take if the sun were the only influence on planetary motion,

is the dominant orbit. The gravitational fields of the planet itself and those of other planets are quite weak compared with that of the sun. By using the observations of Tycho Brahe and observing as carefully and objectively as he could, Kepler plotted the orbits of the planets. Kepler identified the planets' compass direction from his observatory, and their height above the horizon for different times, days, and years. Using the insight he attained from Copernicus, he worked at converting these measures to orbits around the sun. He needed to develop a system that would be consistent with all careful observations of planetary motion, observations made by others as well as himself. It is obvious that if he did not record the time of the measure accurately, or the date, or the compass direction, or the height, or the location on earth from which he measured, he would have anomalous data—that is, data that would not fit a theoretical orbit of a planet. Careful and accurate measurements were essential. Thus, with hard work and insight, Kepler formulated his major laws of planetary motion. The naturally occurring orbital paths are almost identical to Kepler's theoretical ones.

Aristotle, another great scientist, established many of his scientific laws on the basis of casual observation. One of his great laws was that heavy things fall faster than.light ones. This generalization, which had much observation to support it, was accepted as a fact for almost 2000 years. Anyone knows that rocks are heavier than feathers and that they fall faster. In fact, feathers don't even fall; they float. One problem with Aristotle's law is that falling bodies are affected by many things. In a very famous demonstration (probably apocryphal), Galileo went to the top of the Leaning Tower of Pisa, and before a large crowd simultaneously let a heavy ball and a light ball fall. To everyone's (except Galileo's) surprise, the balls stayed next to each other during the fall and landed together. By setting up an artificial situation, Galileo provided data that allowed him to establish an important natural law.

A paradox within science is that artificial, or controlled, situations help scientists understand the natural world. In natural settings many aspects of any two events differ; the scientist sets up artificial situations to make events similar to each other in many ways. Consider the example of falling bodies. Many events influence their fall—the impetus with which they start falling, the things that divert their fall, their shape, the amount of air they displace, as well as (possibly) their weight. Also, the difference in their rates of fall may be slight. If two things fall at different times from different heights, observers may not know which actually fell faster. Even if they do, they may not know whether the height from which they started to fall made any difference. The story is that Galileo contrived the time, place, and condition of the fall; only by doing so could he discover certain natural laws. By dropping two balls of different weight from a great height at the same time, he controlled the height of the fall, the time of the fall, the shape of the objects, and the material of which they were made. By letting the objects fall a long distance at the same time, he could determine even slight differences of speed. When they fell together, he could easily conclude that weight per se does not influence the speed of falling objects. Observations in natural, or uncontrolled, settings had never led to this conclusion.

To study this phenomenon Galileo actually built long, smooth inclined planes and rolled balls down them. He could have watched logs and rocks roll down hills, but again many factors would have been different at different times. Galileo worked for years carefully releasing balls and timing how long it took them to cover different distances. By controlling the situations—that is, making them artificial—he could determine and study the important relations. Like Kepler, he followed false leads and had trouble establishing simple relationships, but he finally came to important conclusions concerning motion. One conclusion was that a constant force does not result in a constant velocity; rather, it results in a constant acceleration. Time and acceleration were seen to be more important variables than velocity and distance. These important discoveries could not have been made if Galileo had not invented artificial situations in order to control and accurately measure the effects of important variables.

Since there are so many things happening in the real world, explaining why and how things happen requires creation of situations in which less is happening. Once laws have been established in simple situations, combining the effects of different variables is usually less difficult. However, even this part of the game may not be won. For example, we know how to describe the path of an object coming near a stationary object, but we still cannot solve the laws of the paths of three moving objects passing near one another.

The use of relatively simple, controlled, artificial situations to study the influence of certain variables in well-defined contexts is particularly necessary in the biological and the social sciences, because the influences of extraneous variables may not be so obvious as in the physical sciences.

An example of the importance of control and artificial situations in biological sciences is the formulation of the genetic theory of heredity by Gregor Mendel and its continuing extension by Barbara McClintock and others. It had long been known that certain traits of one generation are passed to another. However, the explanation of this phenomenon was rather vague and was somehow attributed to the blood of the parents. This is where such expressions as full-blooded and half-blooded come from. Even plants supposedly transmitted traits to their offspring through a fluid in some manner. Both plant and animal hybrids had been grown and bred without the processes involved being understood. Certain individuals had spent many years investigating hybrids and had some knowledge about them—for example, they knew that hybrids tend to return to pure species over generations—but none had a reasonable theory to cover such phenomena.

In 1857 Mendel began to study different varieties of garden peas. He bred them by carefully controlling their pollination for about eight years before reporting his results. His paper furnishes us with an example of careful experimental control and can still be read with profit. (It is reprinted in Newman's World of Mathematics.) In each experiment, Mendel considered for study only two varieties of a single species. In one experiment, these varieties greatly differed in size, one being 6–7 feet tall and the other only 3/4 to 1 1/2 feet tall. He inbred these strains for two generations to confirm that they were stable varieties. This was done by demonstrating that all offspring emulated their parents. He then made a hybrid strain by dusting the pollen from the tall plants onto the eggs of the dwarf plants and vice versa. He covered the plants prior to their dusting so that no other pollination could take place. The resulting hybrid pea plants were all at least as tall as the tall parent plants.

Mendel called the trait that was transmitted to the offspring dominant, and the trait that disappeared recessive. He used the term recessive because the trait that disappeared in the first hybrid generation reappeared in successive generations. For the next generation Mendel self-pollinated all of the plants. That is, he dusted the pollen of a plant onto the plant's own eggs, and then planted the resultant seeds. The new generation included both tall and dwarf pea plants.

About 1/4 of the plants had retained their grandparent's dwarf stature. The other 3/4 of the plants were tall like their parents.

Description of the results of self-fertilization of the plants from this generation is somewhat more complex. The offspring dwarf plants were all dwarf. About 1/3 of the tall plants had offspring that were tall. The other 2/3 of the tall plants had both tall and dwarf offspring in the same mix as their parents did, about 1/4 dwarf and 3/4 tall.

In later generations self-fertilized dwarf plants had only dwarf offspring. Self-fertilized tall plants with only tall siblings had only tall offspring. About 1/3 of the tall plants with both tall and dwarf siblings had only tall offspring, and the rest of these had both tall and dwarf offspring in a ratio of about 3 to 1, respectively.

Later experiments combined plants that differed in more than one factor—for example, the form of the fully developed seed (round or wrinkled), and the color of the seed coat (yellow or green). Mendel concluded that the different traits in hybrid union, as represented in their offspring, were totally independent of one another. He reported that all possible combinations occurred with the mathematical probability identified by random combinations pf dominant and recessive factors. These latter results have been modified by later research. We now believe that, on occasion, different traits are linked to one another.

Mendel proposed that both the pollen and egg cells contain hereditary factors (called genes since about 1909). Each characteristic of an organism is represented in both its pollen and its eggs by one of these hereditary factors. After fertilization, the seed contains the factors from both sources. In constant strains, the genes from the pollen and the egg are the same. In hybrids, corresponding genes represent different traits. Thus, when a tall pea plant and a dwarf one were cross-fertilized the resultant seed contained both factors. In peas, the tall trait is dominant, so the dwarf factor, although present in the seed, could not become manifest in the plant grown from it. This recessive factor, however, has the same chance as the dominant one to be represented in the pollen or egg cells of that plant. In self-pollination, this hybrid randomly combines some pollen containing the tall factor so that some of it combines with eggs containing the tall factor, and some of it combines with eggs containing the dwarf factor. Pollen containing the dwarf factor are also randomly combined with eggs containing each factor. Some of the resultant seeds will contain only the dwarf factor, some will contain only the tall factor, and some (approximately twice as many) will contain both factors. Other factors combine in the same way, and according to Mendel, all such traits were independent of one another.

With well-designed experiments such as the ones Mendel performed, one is in the position to find out just how different aspects of a system affect one another. This kind of specific information aids the development of theories that account for the phenomena and relations studied. It would have been difficult, if not impossible, for Mendel to invent such an insightful theory without having a consistent and constrained data base on which to theorize.

It is important to note, before we leave this section on scientific methodology, that although experimental methods are extremely important, and the most efficient means of establishing functional relationships, they are not the sole means of gaining important scientific information. Many sciences use careful observation as a means of obtaining critical data, from which they may invent and support major theoretical insights.

A critical factor involved in scientific methodology is to define a relatively constrained domain within which to evaluate a meaningful or important question or set of questions. Questions are meaningful to the extent that they are central to the structure or function of the phenomena under study. Scientific research is difficult, in part, because the domain should be a coherent "system" in nature. That is, the things studied should have relatively strong internal relations among one another, and the whole system should often act as a unit among other factors not under study. Sometimes people disagree because it is not obvious where these "joints" in nature are. Major advances often occur when a scientist identifies a set of relationships in one observational context, that can be shown to generalize across many contexts. Mendel, for example, studied the height of pea plants and the color of peas, but he identified some relationships that hold from bacteriophages and fungi through maize, mice, and human beings.

Once the domain is identified, it is important to find an observational context and to observe all of the relevant information within that context. Then the scientist attempts to build or extend a coherent framework that incorporates the collected data as a natural consequence.

Mendel successfully created his theory by identifying a domain of research (successive generations of a single species), carefully constraining the organisms he investigated, and controlling the genetic factors that affected them. His work took place over several years in a corner of his monastery. Darwin took the variability of species as his domain. The question he considered was "What is the origin of the numerous species?" He carefully observed the similarities and differences of species and the environmental niche in which they were successful. He did not control or manipulate his data, but he carefully studied it and compared relations among species from all over the world.

Kepler took as his domain the motion of the planets. His main observational context was the orbit of Mars. His goal was to find a simple regular description that could account for the location of the planets in the heavens at all times. He also observed and compared, but he did not manipulate the phenomena he studied.

We have seen that scientific advancement often depends on artificiality, selection, and control. The scientist, in order to investigate the effects of certain variables, sets up conditions under which other critical variables are not allowed to have an unpredictable effect on the experimental results. By observing the behavior of objects under artificial but well-specified conditions, the scientist can formulate laws that express this behavior. Artificiality and selection, rather than detracting from understanding the events in the real world, create conditions by which one can understand that world. The real world has so many things happening at once that even geniuses such as Galileo or Mendel had to create artificial situations in order to understand natural processes.

Selection and control may be the most powerful methods for attaining knowledge about natural phenomena, but there are many instances in which selection and control seem impossible. Scientists in such areas as archaeology, ethnology, paleontology, political history, sociology, and astronomy cannot manipulate the situation to suit their own purposes. Their only control is to select what and how they observe. These scientists strive to understand phenomena. They invent general principles and attempt to develop a framework within which their data are compatible. For confirmation they try to extend their predictions to include other data. They are limited in that the phenomena are always complex, and they cannot always observe when and what they want. Their task in many ways is much more difficult than that of the experimental scientist, because different events in nature vary in more than one way.

Many sciences present problems that cannot be put into a formal experiment for one reason or another. With some of these problems, it is possible to use statistical techniques to isolate the effects of some of the variables. For example, statistical information confirmed the theory that cigarette smoking leads to lung cancer and certain heart conditions.

No matter what methods scientists use to isolate the effects of variables, their aim is always to achieve a broader understanding of natural processes. And it is this understanding that is the primary goal of science.

The Uncertainty of Science

Although scientists seek understanding, they never achieve it completely. The game of science never ends; all conclusions are tentative. No matter how much information scientists have, they can never be certain of any of their conclusions. There are vast areas in any science where even the working scientists are quite uncertain about how or why events happen the way they do. Most scientists are relatively certain that some currently accepted laws and explanations are quite accurate, but they have no guarantees. The problem is at base a logical one. Philosophers of science currently believe that certainty in the natural sciences is a logical impossibility. They may allow that some areas of endeavor are certain within logic and mathematics, but never in the real world. Even in logic and mathematics, however, there are statements that cannot be evaluated as well as others.

The 18th-century philosopher Immanuel Kant concluded that some things about the world are known for certain—for example, that the sum of the angles of a triangle equals 180° and that all change is continuous. But let's examine these "certainties/' If a triangle is made of wood and the angles are measured as carefully as possible, the angles will not consistently total 180°. Moreover, if a large triangle is conceived as having two stars and the earth as its vertexes, and if straight lines of the sides are measured by light rays (light supposedly travels in a straight line), the angles do not always total the same. In fact, the totals vary significantly. We can conclude from these facts either that light does not always travel in a straight line or that the three angles of a triangle do not always add up to 180°. In neither case are we certain that any physical entity duplicates a Euclidean triangle. We do know that the three angles of a Euclidean triangle add up to 180°, but in practice we can never be certain that we have constructed a Euclidean triangle.

Kant also "knew" for certain that all change is continuous. He meant by this that there cannot be any really sudden changes. You know, for example, that, if you are traveling in a car and suddenly slam on the brakes, the car does not stop immediately; it takes about 210 feet to stop a car traveling at 60 miles per hour. If a rock falls, it gradually increases its speed. Kant assumed that by reason alone he knew that all change is similarly continuous. Even an explosion is supposedly a gradual, though rapid, expansion of gases. We are still not certain whether all change is continuous, but we are certain that we don't know by reason alone that all change is continuous. The question can be debated. In opposition to Kant, current physical theory states that in small systems, such as atoms and molecules, changes are in discrete units. According to this theory there is no such thing as one and a half molecules of water or five and a half photons of light. In addition, if someone turns on a flashlight, the light immediately travels at full speed rather than gradually increasing to maximum speed. Even electrons in their orbits around the nucleus of an atom seem to go from one orbit immediately to another, spending no time getting there. It seems just as easy to believe that all change is discrete as to believe that all change is continuous; we cannot know without experiment which is true. And, since no experiment ever provides absolute certainty (because there are always some uncontrollable variables), we cannot know for certain which is true.

We do have some certain knowledge. For example, the summation 2+2=4 is derivable from definitions. However, we don't know for certain that 2 apples put in a basket with 2 other apples will yield 4 apples. We are pretty sure they will, but we're not absolutely certain.

This is a very difficult concept to grasp, but it is important in understanding the limits of science. There are two ways we can determine with certainty

that a statement is true: (1) we can define it as true, or (2) we can derive it logically from statements that are defined to be true. Since we can't derive any true statements about classes of events in the real world from definitions alone, no scientific statement is true with certainty. For example, we can define the concept material object as something that does not suddenly appear or disappear, and we can define the thing in front of us to be an apple. But we cannot then define the apple to be a material object; we have to investigate to find out whether it is one. And since we have to investigate to find out, we can only know inductively that the apple is a material object, and nothing that is known only by induction is known for certain, as is explained and illustrated below.

If we analyze any law or explanation in natural science, we find that at some level it ultimately rests on induction. In other words, at some level an assumption is made that, since an event has occurred before on several occasions, under similar conditions it will happen again. One reason that no conclusions are certain in science is that there is no way of knowing for certain that the same thing will recur the same way. A low-level example of induction is that we expect the sun to rise every day. We may not see it because of clouds, but we expect it to be there. Why? (1) We have been told that it will rise on a regular schedule, and we may even have had the theory behind that schedule explained to us. (2) We know that it has risen every day of our lives, so why should it stop rising?

Unfortunately, these two reasons are not enough to tell us for certain that the sun will rise tomorrow. (1) Other things we have been told, and have believed, have not always turned out to be true; furthermore, the theories accounting for the things we have been told may be wrong. (2) The evidence that an event has occurred with great regularity is not certain proof that it will continue to do so, as we can see from a short parable. Once upon a time there was a very intelligent turkey. It lived in a pen and was attended by a kind and thoughtful master. All of its desires were taken care of, and it had nothing to do but think of the wonders and regularities of the world. It noticed some major regularities—for example, that mornings always began with the sky getting light, followed by the clop, clop, clop of its master's friendly footsteps. These in turn were always followed by the appearance of delicious food and water within its pen. Other things varied: sometimes it rained and sometimes it snowed; sometimes it was warm and sometimes cold; but amid the variability, footsteps were always followed by food. This sequence was so consistent it became the basis of its philosophy concerning the goodness of the world. One day, after more than 100 confirmations of the turkey's theories, it listened for the clop, clop, clop, heard it, and had its head chopped off. Thus, regularity does not guarantee certainty, and all induction is based on regularity.

Certainty is difficult to attain; not only can we not be certain of the same situation leading to the same results, but in actuality the "same situation" occurs only once. Assume that we are testing the boiling point of water. We pour some distilled water into a beaker, put a thermometer in the beaker, place the beaker in a stand, light the Bunsen burner under it, and wait for the water to boil. After it boils we can read the thermometer. Now assume that we wish to do the same thing the next day. Do we use the same water? It is now a day older. The same beaker? It has been used one more time to boil water. The thermometer also is older and has been used before. If we use different water, a different beaker, and a different thermometer, we do not have the same situation anymore. Furthermore, it is a day later, the sun is up for either a longer or a shorter period of time, we are a day older, the earth has rotated one more time, people have been born and have died, and so on. You may feel that none of this makes a difference; but that is an assumption you make, as does the scientist. The consistency of the data is one way to confirm whether the assumption is essentially correct.

Scientists try to duplicate what they deem to be the significant aspects of a given situation, but each situation is unique, and they do not even attempt to duplicate the total situation. Even the relevant attributes of any situation cannot be duplicated exactly. Scientists accept a certain range of values as duplication. If they are trying for quantitative laws, they will try to limit the range of variation as much as possible, but it cannot be eliminated completely.

We see continued improvement in accuracy of measurement. In all of science, probably the one measure that stands as the most absolute is the speed of light. In late 1972, the accuracy of measurement of that speed was increased one hundredfold by Kenneth Evenson and his research team at the National Bureau of Standards. That research team estimated the speed of light to an accuracy of 0.5 meter per second. In 1983 there was an interesting switch on this topic. Previously, the length of a meter was determined independently, and the speed of light specified by meter length. Now, because the speed of light is considered to be a stable constant, it is used to define the length of a meter. A meter is defined as the distance light travels in 1/299,792,458ths of a second. Any straightedge of that length is one meter long. Obviously we need indirect and inexact methods to make a measuring stick one meter long.

Another constant in an "exact" science is Avogadro's number. This is the number of molecules in one mole (gram-molecular weight) of a substance. The number is given as 6.02486×10^{23} molecules, but there is a 0.0027% error in this estimate. That is a very small error, but it comes to about 16,000,000,000,000,000,000 molecules. That is, Avogadro's number is accurate to about 16 quintillion molecules.

This lack of complete accuracy and certainty is no cause for alarm. It simply means that many of our answers can be improved. There are theoretical limits to the accuracy of our measures, but we have not reached those limits. Besides the limits of accuracy in measurement, the scientific explanations that we currently entertain are not certain to remain intact. However, it is unlikely that many scientific laws will change significantly in the near future, and whatever changes are made will come from individuals who are committed to one of the arenas where the game of science is played. The professionals usually score the points. Successes in the game may eventually be superseded by other successes, but in spite of that possibility they are exhilarating when they occur.

In this chapter we have examined some of the rules and concepts of the game of science. Basically we found that scientists have a systematic method that they use to get information and that they use their understanding of scientific methods to evaluate the work of others. Scientific knowledge is gained solely by the use of scientific methods. These methods are never certain, but they are the best we have.

How We Know

Snow on Cholera

By Martin Goldstein

Introduction:
The Man, The Background

John Snow was born the son of a farmer in York, England, in 1813. At the age of 14 he was apprenticed to a surgeon in Newcastle, who sent him when he was 18 to attend the sufferers of a major outbreak of cholera in the vicinity. In 1838 he passed his examination in London and became a member of the Royal College of Surgeons. He quickly made significant contributions to medical research: he participated in the development of an air pump for administering artificial respiration to newborn children unable to breathe and invented an instrument for performing thoracic surgery. He made major contributions to the new technique of anesthesia, becoming the leading specialist in London in the administration of ether, but switching to the easier-to-use chloroform when his own experimental studies convinced him of its practicality. He administered chloroform to Queen Victoria on the birth of her children, Prince Leopold and Princess Beatrice. His greatest achievement was his study of cholera, which he described in his monograph "On the Mode of Communication of Cholera," one of the classics of scientific method and a fascinating story fascinatingly written. Snow died in 1858, a relatively young man, while at work on a book entitled *On Chloroform and Other Anaesthetics*.

The concept of communicable diseases—that some diseases are transmitted by close contact from the sick to the well—came into being in the Middle Ages. The ancient Greeks were the first to attempt to look at disease scientifically. They rejected the idea of disease as a punishment for sin or as a consequence of witchcraft, and studied instead the relation of diseases to aspects of the natural environment or the way men live, eat, and work. They noted, for example, that it was unhealthy to live near swamps. But in spite of the fact that they suffered from epidemics of various sorts, they somehow missed recognizing that some diseases are contagious.

The prescriptions for isolation and purification described in the Hebrew Bible for physiological processes such as menstruation and for diseases characterized by discharges or skin lesions apparently are based on the idea of the contagiousness of spiritual uncleanliness, of which the physical disease was merely an external symptom. In the Middle Ages, the Church, confronted with a major epidemic of leprosy, revived the biblical practice of isolation of the sick, and the same methods were applied during the outbreak of the Black Death (bubonic plague) in the fourteenth century. By this time the concept of contagion was well established.

It is interesting that the belief that disease was a consequence of evil behavior coexisted with the recognition of contagion for hundreds of years. Attempts to develop treatments for syphilis were opposed on the grounds that syphilis was a punishment for sexual immorality. Cholera was most prevalent among the poor for reasons that will become apparent and there were many who regarded it also as a just punishment for the undeserving and vicious classes of society. A governor of New York State once stated during a cholera epidemic, "… an infinitely wise and just God has seen fit to employ pestilence as one means of scourging the human race for their sins, and it seems to be an appropriate one for the sins of uncleanliness and intemperance. …" The President of New York's Special Medical Council stated at the onset of an epidemic in 1832, "The disease had been confined to the intemperate and the dissolute with but few exceptions." A newspaper report noted, "Every day's experience gives us increased assurance of the safety of the temperate and prudent, who are in circumstances of comfort. The disease is now, more than before, rioting in the haunts of infamy and pollution. A prostitute at 62 Mott Street, who was decking herself before the glass at 1 o'clock yesterday, was carried away in a hearse at half past three o'clock. The broken down constitutions of these miserable creatures perish almost instantly on the attack.. But the business part of our population, in general, appear to be in perfect health and security." A Sunday School newspaper for children explained, *"Drunkards and filthy wicked people of all descriptions are swept away in heaps, as if the Holy God could no longer bear their wickedness, just as we sweep away a mass of filth when it has become so corrupt that we cannot bear it. … The Cholera is not caused by intemperance and filth in themselves, but is a scourge, a rod in the hand of God."* [2]

By the middle of the nineteenth century some of the major communicable diseases had been identified and well differentiated from each other. This was no easy task in itself. For example, a large number of children's diseases have as symptoms fever and a sore throat; these diseases are still not easy to distinguish today. There are many conditions associated with severe diarrhea, for example, cholera, typhoid, dysentery, bacterial food poisoning, noninfectious diseases of the lower intestine such as colitis, and poisoning with certain drugs. In the nineteenth century it had also been demonstrated that some of the contagious diseases could be artificially transmitted by inoculation of small amounts of "morbid matter" taken from the sick. The modes of transmission of particular diseases such as syphilis, intestinal worms, and skin diseases were known. Further, certain types of living organisms had been shown to cause disease directly: the itch mite in scabies as well as certain types of fungus in a disease of silkworms, in ringworm, and in other conditions. Bacteria and protozoa were discovered with the invention and further development of the microscope. These were often observed in the bodies of victims of certain diseases, and various scientists were beginning to speculate that they might be the cause of communicable diseases. The idea in various forms was in the air by the time of Snow, and he made use of it. Solid proof of the germ theory of disease came only in the 1860s and 1870s, after Snow's study, in the work of Pasteur and Koch.

The Disease

Cholera is a bacterial disease characterized by severe diarrhea, vomiting, and muscular cramps. The diarrhea can produce extreme dehydration and collapse; death is frequent, and often occurs within hours after the onset of sickness.

The disease had been known to exist in India since the eighteenth century, and occurs there and in other parts of the world today. In the nineteenth century, as travel between Asia and the West became more common and as the crowding of people in urban centers increased as a result of the industrial revolution, major epidemics occurred in Europe and America. England had epidemics in 1831–32, 1848–49, and 1853–54.

The question of how cholera is transmitted was especially difficult. On the one hand there was good evidence that it could be transmitted by close personal contact. Yet there was equally good evidence that some who had close personal contact with the sick, such as physicians, rarely got it, and that outbreaks could occur at places located at great distances from already existing cases of the disease.

A number of theories were proposed, some of which were too vague to be rationally examined but some of which had solid experimental support. A number of people, including both physicians and uneducated laymen, had blamed the water supply. Snow adopted this theory, but refined it by specifically implicating the excretions of the cholera victims.

Snow's genius lay not so much in hitting on the correct mechanism for the spread of the disease as in providing a beautiful and convincing experimental proof of it; he recognized the importance of the circumstance that, by chance, in a single district of London where an outbreak had occurred, some houses got their drinking water from one source and some from another.

That Snow made the case for his own theory so convincing did not relieve him of the obligation to test alternate theories and show that they did not explain the experimental observations as well as his own. We have collected in one section his discussions of these theories and his arguments against them.

The reader should be aware that Snow's theory did not, at the time he proposed it, explain every single experimental fact. There were some facts that did not fit, and others that were explained as well or better by other theories. Most successful scientific theories, especially when they are new, are in this position, and those who propose them must have the courage and the judgment to put discordant facts aside at times. There are obvious risks in doing this, but we could not advance without it. We will point out from time to

time places where Snow's explanations of discordant facts are shaky.

Introduction to the Study

We have chosen to tell the story as much as possible in Snow's own words, partly because it conveys more of the direct personal experience of making a major scientific discovery, and partly because Snow tells it so well. Page references are to *Snow on Cholera*, a reprint of two of Snow's major monographs, published by the Commonwealth Fund, New York, 1936. All quotations are from "On the Mode of Communication of Cholera," originally published in 1854. All italics are ours. In the selections from the monograph we have mostly followed Snow's order of presentation except in a few cases where logical clarity is achieved by deviating from it.

Snow begins with a brief historical review, following which he cites evidence to show that cholera can be transmitted by close personal contact with the sick:

The History of Cholera

The existence of Asiatic Cholera cannot be distinctly traced back further than the year 1769. Previous to that time the greater part of India was unknown to European medical men; and this is probably the reason why the history of cholera does not extend to a more remote period. It has been proved by various documents, quoted by Mr. Scot, that cholera was prevalent at Madras in the year above mentioned, and that it carried off many thousands of persons in the peninsula of India from that time to 1790. From this period we have very little account of the disease till 1814, although, of course, it might exist in many parts of Asia without coming under the notice of Europeans. …

In 1817, the cholera prevailed with unusual virulence at several places in the Delta of the Ganges; and, as it had not been previously seen by the medical men practising in that part of India, it was thought by them to be a new disease. At this time the cholera began to spread to an extent not before known; and, in the course of seven years, it reached, eastward, to China and the Philippine Islands; southward, to the Mauritius and Bourbon; and to the north-west, as far as Persia and Turkey. Its approach towards our own country, after it entered Europe, was watched with more intense anxiety than its progress in other directions.

It would occupy a long time to give an account of the progress of cholera over different parts of the world, with the devastation it has caused in some places, whilst it has passed lightly over others, or left them untouched; and unless this account could be accompanied with a description of the physical condition of the places, and the habits of the people, which I am unable to give, it would be of little use.

General Observations on Cholera

There are certain circumstances, however, connected with the progress of cholera, which may be stated in a general way. It travels along the great tracks of human intercourse, never going faster than people travel, and generally much more slowly. In extending to a fresh island or continent, it always appears first at a sea-port. It never attacks the crews of ships going from a country free from cholera, to one where the disease is prevailing, till they have entered a port, or had intercourse with the shore. Its exact progress from town to town cannot always be traced; but it has never appeared except where there has been ample opportunity for it to be conveyed by human intercourse. (pp. 1–2) breathing in a poison there should be some evidence of general illness first.

A consideration of the pathology of cholera is capable of indicating to us the manner in which the disease is communicated. If it were ushered in by fever, or any other general constitutional disorder, then we should be furnished with no clue to the way in which the morbid poison enters the system; whether, for instance, by the alimentary canal, by the lungs, or in some other manner, but should be left to determine this point by circumstances unconnected with the pathology of the disease. But from all that I have been able to learn of cholera, both from my own observations and the descriptions of others, I conclude that cholera invariably commences with the affection of the alimentary canal. The disease often proceeds with so little feeling of general illness, that the patient does not consider himself in danger, or even apply for advice, till the malady is far advanced. …

In all the cases of cholera that I have attended, the loss of fluid from the stomach and bowels has been sufficient to account for the collapse, when the previous condition of the patient was taken into account, together with the suddenness of the loss, and the circumstance that the process of absorption appears to be suspended. … (pp. 10–11)

Diseases which are communicated from person to person are caused by some material which passes from the sick to the healthy, and which has the property of increasing and multiplying in the systems of the persons it attacks. In syphilis, small-pox, and vaccinia, we have physical proof of the increase of the morbid material, and in other communicable diseases the evidence of this increase, derived from the fact of their extension, is equally conclusive. As cholera commences with an affection of the alimentary canal, and as we have seen that the blood is not under the influence of any poison in the early stages of this disease, *it follows that the morbid material producing cholera must be introduced into the alimentary canal—must, in fact, be swallowed accidentally, for persons would not take it intentionally; and the increase of the morbid material or cholera poison, must take place in the interior of the stomach and bowels.* It would seem that the cholera poison, when reproduced in sufficient quantity, acts as an irritant on the surface of the stomach and intestines, or, what is still more probable, it withdraws fluid from the blood circulating in the capillaries, by a power analogous to that by which the epithelial cells of the various organs abstract the different secretions in the healthy body. For the morbid matter of cholera having the property of reproducing its own kind, must necessarily have some sort of structure, most likely that of a cell. It is no objection to this view that the structure of the cholera poison cannot be recognized by the microscope, for the matter of small-pox and of chancre can only be recognized by their effects, and not by their physical properties.

The period which intervenes between the time when a morbid poison enters the system, and the commencement of the illness which follows, is called the period of incubation. It is, in reality, a period of reproduction, as regards the morbid matter; and the disease is due to the crop or progeny resulting from the small quantity of poison first introduced. In cholera, this period of incubation or reproduction is much shorter than in most other epidemic or communicable diseases. From the cases previously detailed, it is shown to be in general only from 24 to 48 hours. It is owing to this shortness of the period of incubation, and to the quantity of the morbid poison thrown off in the evacuations, that cholera sometimes spreads with a rapidity unknown in other diseases. ...(pp. 16–17)

Note Snow's speculation, "For the morbid matter of cholera, having the property of reproducing its own kind, must necessarily have some sort of structure, most likely that of a cell." The theory that contagious diseases are caused by microorganisms had been proposed, as noted, by others before Snow, using the same argument. Convincing confirmation did not come for another 20–30 years.

In the first sentence of the next quotation, Snow states the basic hypothesis of his work.

Snow's Theory

The instances in which minute quantities of the ejections and dejections of cholera patients must be swallowed are sufficiently numerous to account for the spread of the disease; and on examination it is found to spread most where the facilities for this mode of communication are greatest.

In the following, Snow points out that people belonging to different social classes perform different functions around the sick, live in different kinds of houses, and have different personal habits and lifestyles. The result is that they have different risks of catching diseases.

Why Doctors Didn't Get Cholera and Those Laying Out the Body Did

Nothing has been found to favour the extension of cholera more than want of personal cleanliness, whether arising from habit or scarcity of water, although the circumstance till lately remained unexplained. The bed linen nearly always becomes wetted by the cholera evacuations, and as these are devoid of the usual colour and odour, the hands of persons waiting on the patient become soiled without their knowing it; and unless these persons are scrupulously cleanly in their habits, and wash their hands before taking food, they must accidentally swallow some of the excretion, and leave some on the food they handle or prepare, which has to be eaten by the rest of the family, who, amongst the working classes, often have to take their meals in the sick room: hence the thousands of instances in which, amongst their class of the population, a case of cholera in one member of the family is followed by other cases; whilst medical men and others, who merely visit the patients, generally escape. The post mortem inspection of the bodies of cholera patients has hardly ever been followed by the disease that I am aware, this being a duty that is necessarily followed by careful washing of the hands; and it is not the habit of medical men to be taking food on such an occasion. On the other hand, the duties performed about the body, such as laying it out, when done by women of the working class, who make the occasion one of eating and drinking, are often followed by an attack of cholera; and persons who merely attend the funeral, and have no connexion with the body, frequently contract the disease, in consequence, apparently, of partaking of food which has been prepared or handled by those having duties about the cholera patient, or his linen and bedding. ... (pp. 16–17)

Why the Rich Did Not Get Cholera So Often

The involuntary passage of the evacuations in most bad cases of cholera, must also aid in spreading the disease. Mr. Baker, of Staines, who attended 260 cases of cholera and diarrhea in 1849, chiefly among the poor, informed me that "when the patients passed their stools involuntarily the disease evidently spread." It is amongst the poor, where a whole family live, sleep, cook, eat and wash in a single room, that cholera has been found to spread when once introduced, and still more in those places termed common lodging-houses, in which several families were crowded into a single room. It was amongst the vagrant class, who lived in this crowded state, that cholera was most fatal in 1832; but the Act of Parliament for the regulation of common lodging-houses, has caused the disease to be much less fatal amongst these people in the late epidemics. When, on the other hand, cholera is introduced into the better kind of houses, as it often is, by means that will be afterwards pointed out, it hardly ever spreads from one member of the family to another. The constant use of the hand-basin and towel, and the fact of the apartments for cooking and eating being distinct from the sick room, are the cause of this. ... (p. 18)

We may recall the two observations cited by Snow against the "effluvia" hypothesis: (1) not everyone (such as doctors) having close contact with a cholera victim gets it, and (2) sometimes it appears at great distances from the nearest case. Note that Snow's hypothesis (communication through evacuation) explains the first observation plausibly. Now he deals with the second, by making an additional subsidiary hypothesis.

How Did Cholera Get to the Rich?

If the cholera had no other means of communication than those which we have been considering, it would be constrained to confine itself chiefly to the crowded dwellings of the poor, and would be continually liable to die out accidentally in a place, for want of the opportunity to reach fresh victims; but there is often a way open for it to extend itself more widely, and to reach the well-to-do classes of the community; I allude to the mixture of the cholera evacuations with the water used for drinking and culinary purposes, either by permeating the ground, and getting into wells, or by running along channels and sewers into the rivers from which entire towns are sometimes supplied with water. ... (pp. 22–23)

In the following quotations Snow gives evidence for his second hypothesis: *that cholera spreads through the water supply.* It should be noted that the idea that water is responsible was suggested by many others quoted by Snow: Mr. Grant, Dr. Chambers, Mr. Cruikshanks, and various other named and unnamed individuals. Snow's hypothesis, although it must have owed much to these people, is more specific in that it identifies the excretions of the victims as the source of the contamination of the water supply, and explains transmission by direct contact as well.

In 1849 there were in Thomas Street, Horsleydown, two courts close together, consisting of a number of small houses or cottages, inhabited by poor people. The houses occupied one side of each court or alley—the south side of Trusscott's Court, and the north side of the other, which was called Surrey Buildings, being placed back to back, with an intervening space, divided into small back areas, in which were situated the privies of both the courts, communicating with the same drain, and there was an open sewer which passed the further end of both courts.

> Now, in Surrey Buildings the cholera committed fearful devastation, whilst in the

adjoining court there was but one fatal case, and another case that ended in recovery. In the former court, the slops of dirty water, poured down by the inhabitants into a channel in front of the houses, got into the well from which they obtained their water; this being the only difference that Mr. Grant, the Assistant-Surveyor for the Commissioners of Sewers, could find between the circumstances of the two courts as he stated in a report that he made to the Commissioners. ... (p. 23)

In Manchester, a sudden and violent outbreak of cholera occurred in Hope Street, Salford. The inhabitants used water from a particular pump-well. This well had been repaired, and a sewer which passes within 9 inches of the edge of it became accidentally stopped up, and leaked into the well. The inhabitants of 30 houses used the water from this well; among them there occurred 19 cases of diarrhea, 26 cases of cholera, and 25 deaths. The inhabitants of 60 houses in the same immediate neighbourhood used other water; among these there occurred 11 cases of diarrhaea, but not a single case of cholera, nor one death. It is remarkable, that, in this instance, out of the 26 persons attacked with cholera, the whole perished except one. ... (pp. 31–32)

The Washerwoman Was Spared

Dr. Thomas King Chambers informed me, that at Ilford, in Essex, in the summer of 1849, the cholera prevailed very severely in a row of houses a little way from the main part of the town. It had visited every house in the row but one. The refuse which overflowed from the privies and a pigsty could be seen running into the well over the surface of the ground, and the water was very fetid; yet it was used by the people in all the houses except that which had escaped cholera. That house was inhabited by a woman who took linen to wash, and she, finding that the water gave the linen

an offensive smell, paid a person to fetch water for her from the pump in the town, and this water she used for culinary purposes, as well as for washing.

How the Landlord Got It

The following circumstance was related to me, at the time it occurred, by a gentleman well acquainted with all the particulars. The drainage from the cesspools found its way into the well attached to some houses at Locksbrook, near Bath, and the cholera making its appearance there in the autumn of 1849, became very fatal. The people complained of the water to the gentleman belonging to the property, who lived at Weston, in Bath, and he sent a surveyor, who reported that nothing was the matter. The tenants still complaining, *the owner went himself, and on looking at the water and smelling it, he said that he could perceive nothing the matter with it.* He was asked if he would taste it, *and he drank a glass of it. This occurred on a Wednesday; he went home, was taken ill with the cholera, and died on the Saturday following, there being no cholera in his own neighbourhood at the time.* ... (pp. 31–32)

The First Experiment: 1849

The Broad Street Pump

By the time of the 1849 outbreak of cholera in the vicinity of the Broad Street pump, described next, Snow already was convinced that cholera is spread through the water supply, but the data he was able to gather by close observation guided by his hypothesis made the case for it much more convincing. Because of his hypothesis, he asked certain questions and noticed certain things, for example, the high rate of the disease among the customers of a certain coffee shop, and the low rate among the inhabitants of a workhouse and the employees at a brewery.

He also took the first public health measure based on his ideas. He told the Board of Guardians of the parish to remove the handle of the Broad Street pump to prevent any further use of the contaminated water and thus any further cases of cholera arising from this source. He hoped that this would provide experimental proof of his theory. It would have done so if there had been a sudden drop in the number of new cases of the disease after the pump handle was removed. But in this he was disappointed. The epidemic had already passed its peak and the number of new cases was already falling rapidly.

"The Mortality in This Limited Area Equals Any That Was Ever Caused in This Country by the Plague"

The most terrible outbreak of cholera which ever occurred in this kingdom is probably that which took place in Broad Street, Golden Square, and the adjoining streets, a few weeks ago. Within 250 yards of the spot where Cambridge Street joins Broad Street, there were upwards of 500 fatal attacks of cholera in 10 days. The mortality in this limited area probably equals any that was ever caused in this country, even by the plague; and it was much more sudden, as the greater number of cases terminated in a few hours. The mortality would undoubtedly have been much greater had it not been for the flight of the population. Persons in furnished lodgings left first, then other lodgers went away, leaving the furniture to be sent for when they could meet with a place to put it in. Many houses were closed altogether, owing to the death of the proprietors; and, in a great number of instances, the tradesmen who remained had sent away their families so that in less than six days from the commencement of the outbreak, the most afflicted streets were deserted by more than three-quarters of their inhabitants.

There were a few cases of cholera in the neighbourhood of Broad Street, Golden Square, in the latter part of August; and the so-called outbreak which commenced in the night between the 31st August and the 1st September, was, as in all similar instances, only a violent increase of the malady. As soon as I became acquainted with the situation and extent of this irruption of cholera, I suspected some contamination of the water of the much-frequented street-pump in Broad Street, near the end of Cambridge Street; but on examining the water, on the evening of the 3rd September, I found so little impurity in it of an organic nature, that I hesitated to come to a conclusion. Further inquiry, however, showed me that there was no other circumstance or agent common to the circumscribed locality in which this sudden increase of cholera occurred, and not extending beyond it, except the water of the above mentioned pump. ... (pp. 38–39)

Snow began his study by obtaining from the London General Register Office a list of the deaths from cholera in the area occurring each day. These figures showed a dramatic increase in cases on August 31, which he therefore identified as the starting date of the outbreak. He found 83 deaths that took place from August 31 to September 1 (see Table 1), and made a personal investigation of these cases.

On proceeding to the spot, I found that nearly all the deaths had taken place within a short distance of the pump. There were only ten deaths in houses situated decidedly nearer to another street pump. In five of these cases the families of the deceased persons informed me that they always sent to the pump in Broad Street, as they preferred the water to that of the pump which was nearer. In three other cases, the deceased were children who went to school near the pump in Broad Street. Two of them were known to drink the water; and the parents of the third think it probable that it did so. The other two deaths, beyond the district which this pump supplies, represent only the amount of mortality from cholera that was occurring before the irruption took place.

With regard to the deaths occurring in the locality belonging to the pump, there were 61 instances in which I was informed that the deceased persons used to drink the pump-water from Broad Street, either constantly or occasionally. In six instances I could get no information, owing to the death or departure of everyone connected with the deceased individuals; and in six cases I was informed that the deceased persons did not drink the pump-water before their illness. ... (pp. 39–40)

Who Drank the Pump Water?

For reasons of clarity we summarize the results of Snow's investigation of these 83 deaths in Table 1, which shows that there were deaths among people not known to have drunk water from the Broad Street pump. These deaths therefore are facts that seem to contradict Snow's hypothesis. A scientist faced with facts contradictory to a hypothesis has many alternatives, only one of which is to discard the hypothesis. Another alternative is to make a closer examination of these facts, to see whether in some plausible way they can be shown either not really to contradict the hypothesis or actually to support it. It occurred to Snow to look for ways *the individuals in question might have drunk the water without being aware of it.*

The additional facts that I have been able to ascertain are in accordance with those above related; and as regards the small number of those attacked, who were believed not to have drunk the water from Broad Street pump, it must be obvious that there are various ways in which the deceased persons may have taken it without the knowledge of their friends. The water was used for mixing with spirits in all the public houses around. It was used likewise at dining-rooms and coffee-shops. The keeper of a coffee-shop in the neighbourhood, which was frequented by mechanics, and where the pump-water was supplied at dinner time, informed me (on 6th September) that she was already aware of nine of her customers who were dead. The pump-water was also sold in various little shops, with a teaspoonful of effervescing powder in it, under the name of sherbet; and it may have been distributed in various other ways with which I am unacquainted. The pump was frequented much more than is usual, even for a London pump in a populous neighbourhood, (pp. 41–42)

Snow next gives two striking observations that confirm the role of the pump. There were two large groups of people living near the Broad Street pump who had very few cases of cholera: the inhabitants of a workhouse and the employees of a brewery.

TABLE I Results of Snow's Investigation

83 Deaths[a]					
73 Living near Broad Street pump			10 Not living near pump		
61 Known to have drunk pump water	6 Believed not to have drunk pump water	6 No information	5 In families sending to Broad St. pump for water	3 Children attending school near pump	2 No information

"Out of 83 individuals who had died of the disease, 69 were known definitely or could be assumed to have drunk the pump water, 6 were believed not to have drunk it, and for 8 there was no information.

Why Were the Workhouse and the Brewery Spared?

In the workhouse, which had its own water supply, only 5 out of 535 inmates died. If the death rate had been the same as in the surrounding neighborhood, over 100 would have died.

There is a brewery in Broad Street, near to the pump, and on perceiving that no brewer's men were registered as having died of cholera, I called on Mr. Huggins, the proprietor. He informed me that there were about 70 workmen employed in the brewery, and that none of them had suffered from cholera—at least in a severe form—only two having been indisposed, and that not seriously, at the time the disease prevailed. The men are allowed a certain quantity of malt liquor, and Mr. Huggins believes that they do not drink water at all; and he is quite certain that the *workmen never obtained water from the pump in the street. There is a deep well in the brawery,* in addition to the New River water. ... (pp. 41–42)

The Pump Handle

On September 7, Snow met with the Board of Guardians of the parish and informed them of his evidence as to the role of the pump in the outbreak. On September 8, the handle of the pump was removed, but, as Snow notes, by this time the epidemic had subsided, perhaps because many inhabitants had fled the neighborhood. So the removal of the pump handle did not produce any dramatic effect on the number of new cases.

Following the epidemic, the pump was opened and examined. No direct evidence of leakage from nearby privies was found, but Snow states his belief that it must have occurred, perhaps by seepage through the soil, as on microscopic examination "oval animalcules" were found, which Snow points out are evidence of organic contamination. (They were not the bacteria causing cholera, which were not detectable by the microscopic techniques of the time, nor

did Snow take them seriously as a causative agent—rather, he knew that "animalcules" were very common in natural waters contaminated with sewage or other organic matter, even when no cholera was present.)

Additional evidence for the contamination of the pump water with sewage was provided by inhabitants of the neighborhood who had noticed a disagreeable taste in the water just prior to the outbreak and a tendency of the water to form a scum on the surface when it was left to stand a few days. Further, chemical tests showed the presence of large amounts of chlorides, consistent with contamination by sewage, but, like the animalcules, not constituting overwhelming proof. The question of chlorides in the drinking water will come up again more dramatically later on.

Snow's conclusion on the Broad Street pump outbreak is as follows:

> Whilst the presumed contamination of the water of the Broad Street pump with the evacuations of cholera patients affords an exact explanation of the fearful outbreak of cholera in St. James's parish, there is no other circumstance which offers any explanation at all, whatever hypothesis of the nature and cause of the malady be adopted. .... (pp. 51–54)

The Second Experiment: 1853–54

A Controlled Experiment—Where Did They Get Their Water?

The next section is the heart of Snow's monograph. It describes his observations during the 1853–54 outbreak and his performance of a controlled experiment to test his theory.

The basic idea of a controlled experiment is simple. Suppose mice could get cholera, and one wished to prove that water containing the excretions of cholera victims could produce the disease. One would take two large groups of mice similar in every relevant respect, put cholera excretions in the drinking water of one group (the test group), and leave them out of the water of the second group (the control group). If a large number of cholera cases were found in the test group and none in the control group, the case would have been made.

Human beings are more difficult to experiment on than mice. First, there is an ethical question—if you are inclined to believe that contaminated drinking water produces cholera, even though you are not yet sure, do you have the right to let people drink it? Even if they would be drinking it anyway, don't you have an obligation to stop them?

The ethical problems can be avoided if by chance a "natural" experiment is available: it may happen that a group in the population has been exposed fortuitously to what is believed to be the cause of a disease. A controlled experiment is then possible if another group in the population can be found, similar in every relevant respect to the first one, except that it has not been exposed to the suspected cause. If the disease occurs in the first group and not the second, we have confirming evidence that the suspected cause really is the cause. But in such a "natural" situation it may be hard to prove the two groups similar in "every relevant respect."

For example, different districts of London had different water-supplies and different cholera rates. But unfortunately, from the point of view of testing Snow's hypothesis that cholera is caused by contaminated water, the people in the different districts were different in other ways, also. The rich lived in different neighborhoods from the poor and suffered less from cholera. Was it because they had uncontaminated water supplies or because they ate better food, worked shorter hours at easier jobs, lived in newer, cleaner houses?

Also, different groups of equally "poor" people might differ in other significant ways. In London at that time there was a tendency for people of the same occupation to live in a single neighborhood, so that one neighborhood might have a lot of butchers, another might have tailors, and a third drivers of carts. Might susceptibility to cholera depend on occupation? Snow himself was aware that some occupational groups such as doctors were less likely to get cholera, and some, such as coal miners, were more likely. Perhaps some overlooked causative factor was related to one's work.

Since we now know that Snow's theory about the water supply was correct, we can feel that all these other differences are irrelevant and can be disregarded. But at the time this wasn't yet clear, and of course the purpose of the experiment was to find this out. If the control and test groups differed in three or four other ways besides getting their water supplies from a different source, we would not feel safe in blaming the water supply alone; any of these other differences between the groups might be responsible for the differences in cholera rates.

The Natural Experiment

It was Snow's genius to recognize the importance of the fortuitous circumstance that two different water companies supplied a single neighborhood in an intermingled way.

The two water companies in question both drew their water from the Thames, from spots that could be expected to be contaminated with the sewage of the city. But in 1852, after the epidemic during which Snow had done the experiments described above, one of these companies, the Lambeth Company, moved their waterworks upstream to a place free of London sewage. The other, the Southwark and Vauxhall Company, remained where it was. Both companies delivered drinking water to a single district of the city:

> The pipes of each Company go down all the streets, and into nearly all the courts and alleys. A few houses are supplied by one Company and a few by the other, according to the decision of the owner

or occupier at that time when the Water Companies were in active competition. In many cases a single house has a supply different from that on either side. Each Company supplies both rich and poor, both large houses and small; *there is no difference either in the condition or occupation of the persons receiving the water of the different Companies.*

In the next sentence, Snow summarizes the basic idea of the experiment:

As there is no difference whatever, either in the houses or the people receiving the supply of the two Water Companies, or in any of the physical conditions with which they are surrounded, it is obvious that no experiment could have been devised which would more thoroughly test the effect of water supply on the progress of cholera than this, which circumstances placed ready-made before the observer.

The experiment, too, was on the grandest scale. No fewer than 300,000 people of both sexes, of every age and occupation, and of every rank and station, from gentlefolks down to the very poor, were divided into two groups without their choice, and, in most cases, without their knowledge; one group being supplied with water containing the sewage of London, and, amongst it, whatever might have come from the cholera patients, the other group having water quite free from such impurity.

To turn this grand experiment to account, all that was required was to learn the supply of water to each individual house where a fatal attack of cholera might occur. I regret that, in the short days at the latter part of last year, I could not spare the time

to make the inquiry; and, indeed, I was not fully aware, at that time, of the very intimate mixture of the supply of the two Water Companies, and the consequently important nature of the desired inquiry, (pp. 75–76)

Carrying out the idea required putting together two kinds of data: cholera cases and water supply. The first was easier to come by than the second.

When the cholera returned to London in July of the present year, however, I resolved to spare no exertion which might be necessary to ascertain the exact effect of the water supply on the progress of the epidemic, in the places where all the circumstances were so happily adapted for the inquiry. I was desirous of making the investigation myself, in order that I might have the most satisfactory proof of the truth or fallacy of the doctrine which I had been advocating for 5 years. I had no reason to doubt the correctness of the conclusions I had drawn from the great number of facts already in my possession, *but I felt that the circumstance of the cholera-poison passing down the sewers into a great river, and being distributed through miles of pipes, and yet producing its specific effects, was a fact of so startling a nature, and of so vast importance to the community, that it could not be too rigidly examined, or established on too firm a basis,* (p. 76)

Snow began to gather data on cholera deaths in the district. The very first results were supportive of his conjecture: of 44 deaths in the district in question, 38 occurred in houses supplied by the Southwark and Vauxhall Company.

As soon as I had ascertained these particulars I communicated them to Dr. Farr, who was much struck with the result, and at his suggestion the Registrars of all the south districts of London were requested to make a return of the water supply of the house in which the attack took place, in all cases of death from cholera. This order was to take place after the 26th August, and I resolved to carry my inquiry down to that date, so that the facts might be ascertained for the whole course of the epidemic. ... (p. 77)

Chlorides and Receipts

Determining which water company supplied a given house was not always straightforward. Fortunately, Snow found a chemical test based on the fact that when a solution of silver nitrate is added to water containing chlorides a white cloud of insoluble silver chloride is formed. He found that the water from the two companies differed markedly in chloride content and thus could be easily distinguished.

The inquiry was necessarily attended with a good deal of trouble. There were very few instances in which I could at once get the information I required. Even when the water-rates are paid by the residents, they can seldom remember the name of the Water Company till they have looked for the receipt. In the case of working people who pay weekly rents, the rates are invariably paid by the landlord or his agent, who often lives at a distance, and the residents know nothing about the matter. It would, indeed, have been almost impossible for me to complete the inquiry, if I had not found that I could distinguish the water of the two companies with perfect certainty by a chemical test. The test I employed was found on the great difference in the chloride of sodium contained in the two kinds of water, at the time I made the inquiry. ... (pp. 77–78)

[T]he difference in appearance on adding nitrate of silver to the two kinds of water was so great, that they could be at once distinguished without further trouble. Therefore when the resident could not give clear and conclusive evidence about the water Company, I obtained some of the water in a small phial, and wrote the address on the cover, when I could examine it after coming home. The mere appearance of the water generally afforded a very good indication of its source, especially if it was observed as it came in, before it had entered the water-butt or cistern; and the time of its coming in also afforded some evidence of the kind of water, after I had ascertained the hours when the turncocks of both Companies visited any street. These points were, however, not relied on, except as corroborating more decisive proof, such as the chemical test, or the Company's receipt for the rates. ... (p. 78)

It is worth noting how careful Snow was to be sure of the facts here—although he could guess the source of the water from its "mere appearance," he relied on more objective proof of its origin.

Deaths and Death Rates

Snow now expresses the result of his study in quantitative terms. He notes that the Southwark and Vauxhall Company supplied about 40,000 houses in London during 1853 and the Lambeth Company (drawing its water upstream) about 26,000. In the rest of London, where there were over 250,000 houses, there were *more* deaths than in the houses supplied by Southwark and Vauxhall—1422 compared with 1263—but there were 6 times as many houses, also. What matters here is not the total number of deaths, but the *rate* of deaths per house. Put another way, if you live in a house supplied by Southwark and Vauxhall, what are your chances of dying, compared with your chances if you live in a house supplied by another company? Snow expressed the rate in deaths per 10,000 houses, according to the following formula:

$$\text{Rate} = (\text{Death/number of houses}) \times 10{,}000$$

The mortality in the houses supplied by the Southwark and Vauxhall Company was therefore between eight and nine times as great as in the houses supplied by the Lambeth Company____(p. 86)

Being Critical

Objections to Snow's Theory

Snow next considers an objection to his hypothesis: not everyone who drinks the polluted water gets sick. Note that he had used a similar objection against the "effluvia" hypothesis: not everyone exposed to the effluvia of cholera patients gets sick. However, he dealt differently with the two cases. He was able to find a relevant factor consistent with his own hypothesis to separate those who became ill from those who did not: they were members of different social groups with different sanitary practices that caused them to have different chances of ingesting excreta. On the other hand, Snow did not find, among those with equal chances of ingesting excreta, any factor that distinguished those who became ill from those who did not. Those who did not accept Snow's hypothesis, and had cited as evidence against it the fact that not all known to ingest excreta got the disease, would have been able to make a better case if they had been able to identify a factor distinguishing those who became ill from those who did not that was consistent with an alternative hypothesis.

Here is Snow's discussion of this problem:

> All the evidence proving the communication of cholera through the medium of water, confirms that with which I set out, of its communication in the crowded habitations of the poor, in coal-mines and other places, by the hands getting soiled with the evacuations of the patients, and by small quantities of these evacuations being swallowed with the food, as paint is swallowed by house painters of uncleanly habits, who contract lead-colic in this way.

Why Some Who Should Get Cholera Don't

There are one or two objections to the mode of communication of cholera which I am endeavouring to establish, that deserve to be noticed. Messrs. Pearse and Marston state, in their account of the cases of cholera treated at the Newcastle Dispensary in 1853, that one of the dispensers drank by mistake some rice-water evacuation without any effect whatever. In rejoinder to this negative incident, it may be remarked, that several conditions may be requisite to the communication of cholera with which we are as yet unacquainted. Certain conditions we know to be requisite to the communication of other diseases. Syphilis we know is only communicable in its primary stage, and vaccine lymph must be removed at a particular time to produce its proper effects. In the incident above mentioned, the large quantity of the evacuation taken might even prevent its action. It must be remembered that the effects of a morbid poison are never due to what first enters the system, but to the crop or progeny produced from this during a period of reproduction, termed the period of incubation; and if a whole sack of grain, or seed of any kind, were put into a hole in the ground, it is very doubtful whether any crop whatever would be produced.

An objection that has repeatedly been made to the propagation of cholera through the medium of water, is, that every one who drinks of the water ought to have the disease at once. This objection arises from mistaking the department of science to which the communication of cholera belongs, and looking on it as a question of chemistry, instead of one of natural history, as it undoubtedly is. It cannot be supposed that a morbid poison, which has the property, under suitable circumstances, of reproducing its kind, should be capable of being diluted indefinitely in water, like a chemical salt; and therefore it is not to be presumed that the cholera-poison would be equally diffused through every particle of the water. The eggs of the tape-worm must undoubtedly pass down the sewers into the Thames, but it by no means follows

Table IX

	No. of houses	Deaths from Cholera	Deaths in each 10,000 houses
Southwark & Vauxhall Company	40,046	1263	315
Lambeth Company	26,107	98	37
Rest of London	256,423	1422	59

*Deaths per 10,000 houses.

that everybody who drinks a glass of the water should swallow one of the eggs. As regards the morbid matter of cholera, many other circumstances, besides the quantity of it which is present in a river at different periods of the epidemic must influence the chances of its being swallowed, such as its remaining in a butt or other vessel till it is decomposed or devoured by animalcules, or its merely settling to the bottom and remaining there. In the case of the pump-well in Broad Street, Golden Square, if the cholera-poison was contained in the minute whitish flocculi visible on close inspection to the naked eye, some persons might drink of the water without taking any, as they soon settled to the bottom of the vessel. ... (pp. 111–113)

In some respects Snow's defense against this objection would be accepted as valid today, in the light of knowledge gained in the century that has elapsed since the germ theory of disease was accepted. We know, for example, that individual susceptibilities to a given disease vary widely, often for reasons that even now are not well understood. Also, some individuals may suffer an attack of a disease in a mild and clinically unrecognized form and may subsequently be immune for a longer or shorter period of time. It is a very rare epidemic in which everyone gets sick.

Snow's explanation of why the individual who drank cholera evacuation by mistake did not contract the disease is not plausible today, nor can we believe that it would have been plausible at the time, especially to anyone skeptical of Snow's theory. It would

have been more admirable, but less human, if Snow had acknowledged that this was one experimental fact he couldn't explain, and let it go at that.

Scotland Is Different

The next section discusses one oddity of the behavior of cholera: it was mainly a summer disease in England and would not spread in winter even when introduced then, but it seemed not to be seasonal in Scotland, running through its epidemic course as soon as it appeared, even in winter. Snow's explanation in terms of his theory is charming, and it shows the kinds of things a scientist has to be alert to.

It also may help to demolish a myth about the scientific method we have referred to in an earlier chapter: that scientific hypotheses are obtained by first examining the facts. In reality, the hypothesis comes first, and tells us which facts are worth examining. It is easy to see how Snow was led to compare the drinking habits of the English with those of the Scots, given his theory, but if one had only the facts about the seasonal differences in cholera between the two countries, would one have inferred a theory blaming the water supply?

In the Winter the English Drank Tea

Each time when cholera has been introduced into England in the autumn, it has made but little progress, and has lingered rather than flourished during the winter and spring, to increase gradually during the following summer, reach its climax at the latter part of summer, and decline somewhat rapidly as the cool days of autumn set in. In most parts of Scotland, on the contrary, cholera has each time run through its course in the winter immediately following its introduction. I have now to offer what I consider an explanation, to a great extent, of these peculiarities in the progress of cholera. The English people, as a general rule, do not drink much unboiled water, except in warm weather. They generally take tea, coffee, malt liquor, or some other artificial beverage at their meals, and do not require to drink between meals, except when the weather is warm. In summer, however, a much greater quantity of drink is required, and it is much more usual to drink water at that season than in cold weather. Consequently, whilst the cholera is chiefly confined in winter to the crowded families of the poor, and to the mining population, who, as was before explained, eat each other's excrement at all times, it gains access as summer advances to the population of the towns, where there is a river which receives the sewers and supplies the drinking water at the same time; and, where pump-wells and other limited supplies of water happen to be contaminated with the contents of the drains and cesspools, there is a greater opportunity for the disease to spread at a time when unboiled water is more freely used.

While the Scots ...

In Scotland, on the other hand, unboiled water is somewhat freely used at all times to mix with spirits; I am told that when two or three people enter a tavern in Scotland and ask for a gill of whiskey, a jug of water and tumbler-glasses are brought with it. Malt liquors are only consumed to a limited extent in Scotland, and when persons drink spirit without water, as they often do, it occasions thirst and obliges them to drink water afterwards, (pp. 117–118)

Other Theories: Effluvia, Elevation, Hard Water, and Soft Water

We have collected in one place Snow's discussion of alternate theories and his reasons for rejecting them. Giving fair consideration to theories opposed to one's own is something all scientists should try to do; but not all scientists are really capable of it, and are not necessarily bad scientists because of this shortcoming. Science proceeds by a consensus of scientists: one man's failure to be objective about a theory he doesn't like is made up for by the opposite bias of his opponents and the fairness of the less emotionally involved. Snow was better at it than most.

Whilst the presumed contamination of the water of the Broad Street pump with the evacuations of cholera patients affords an exact explanation of the fearful outbreak of cholera in St. James's parish, there is no other circumstance which offers any explanation at all, whatever hypothesis of the nature and cause of the malady be adopted. ... Many of the non-medical public were disposed to attribute the outbreak of cholera to the supposed existence of a pit in which persons dying of the plague had been buried about two centuries ago; and, if the alleged plague-pit had been nearer to Broad Street, they would no doubt still cling to the idea. The situation of the supposed pit is, however, said to be Little Marlborough Street, just out of the area in which the chief mortality occurred. With regard to effluvia from the sewers passing into the streets and houses, that is a fault common to most parts of London and other towns. There is nothing peculiar in the sewers or drainage of the limited spot in which this outbreak occurred; and Saffron Hill and other localities, which suffer much more from ill odours, have been very lightly visited by cholera. ... (pp. 54–55)

The low rate of mortality amongst medical men and undertakers is worthy of notice. If cholera were propagated by effluvia given off from the patient, or the dead body, as used to be the opinion of those who believed in its communicability; or, if it depended on effluvia lurking about what are by others called infected localities, in either case medical men and undertakers would be peculiarly liable to the disease; but, according to the principles explained in this treatise, there is no reason why these callings should particularly expose persons to the malady, (p. 122)

It is easy today to look down on the effluvia theory as so much unenlightened superstition. One should recognize, however, that the germ theory then was highly speculative and had very little evidence in its favor. The idea that disease could be spread by foul odors or other poisonous emanations represented a great advance over views attributing disease to witchcraft or sin, and, in the absence of any knowledge of microorganisms, was a plausible explanation of contagion.

Further, the effluvia theory led to justified concern over the crowded and unsanitary living and working conditions of the poor. Interested readers should consult the report prepared for Parliament by E. Chadwick in 1842 for a description (of these conditions.[3] Chadwick's report led to the first serious public health measures taken by the British government, and in fact these measures resulted in improved health of the population of England.

This illustrates a truism of scientific research: an incorrect theory is better than no theory at all, or, in the words of an English logician Augustus de Morgan, "Wrong hypotheses, rightly worked, have produced more useful results than unguided observation. "[4]

Height Above Sea Level

Dr. Farr's theory that cholera is less prevalent in a district the higher its elevation above sea level is treated differently from the effluvia theory by Snow. The latter he rejects completely, but Farr's theory had some validity at least within London. Snow shows that the limited correlation between the disease and elevation noted by Farr is actually better explained by his own theory: the low-lying districts of London are also those more likely to have water supplies contaminated by sewage.

Farr's observations can be thought of as a controlled experiment, the control and test groups being inhabitants of London living at different elevations above sea level. Indeed, the people living at the lower elevations suffered more cholera, but the two groups differed, as pointed out by Snow, in other significant ways, even though elevation was directly connected to the differences in the significant factor.

Dr. Farr discovered a remarkable coincidence between the mortality from cholera in the different districts of London in 1849, and the elevation of the ground; the connection being of an inverse kind, the higher districts suffering least, and the lowest suffering most from this malady. Dr. Farr was inclined to think that the level of the soil had some direct influence over the prevalence of cholera, but the fact of the most elevated towns in this kingdom, as Wolverhampton, Dowlais, Merthyr Tydvil, and Newcastle-upon-Tyne, having suffered excessively from this disease on several occasions, is opposed to this view, as is also the circumstance of Bethlehem Hospital, the Queen's Prison, Horsemonger Lane Gaol, and several other large buildings, which are supplied with water from deep wells on the premises, having nearly or altogether escaped cholera, though situated on a very low level, and surrounded by the disease. The fact of Brixton, at an elevation 56 feet above Trinity high-water mark, having suffered a mortality of 55 in 10,000, whilst many districts on the north of the Thames, at less than half the elevation, did not suffer one-third as much, also points to the same conclusion.

I expressed the opinion in 1849, that the increased prevalence of cholera in the low-lying districts of London depended entirely on the greater contamination of the water in these districts, and the comparative immunity from this disease of the population receiving the improved water from Thames Ditton, during the epidemics of last year and the present, as shown in the previous pages, entirely confirms this view of the subject; for the great bulk of this population live in the lowest districts of the metropolis . ... (pp. 97–98)

Limestone and Sandstone

Another hypothesis, which agreed with at least some of the experimental facts, was proposed by John Lea of Cincinnati. Lea had found that districts in which the underlying rock formations were limestone had much more cholera than districts overlying sandstone. He conjectured that the calcium and magnesium salts, which were present in water in limestone districts, were somehow necessary for the cholera "poison" to have its effect. He noted as supporting evidence for this hypothesis the fact that towns that relied on river water, in which there was much calcium and magnesium, suffered more than towns that used rain water.

Snow's criticisms of Lea's hypothesis is in part specious. He attributed the difference in cholera rates between sandstone and limestone districts observed by Lea to a greater oxidizing power or sandstone on organic substances. This explanation is not very plausible, as Snow himself was aware—he had no evidence that limestone might not be equally oxidizing. We can be even more sure today that the correlation between cholera and rock formation found by Lea was entirely fortuitous. Snow of course explained the higher cholera rates in towns using river water on the greater likelihood that river water is contaminated with sewage.

Applications to Other Problems

What About Other Diseases?

Having established convincingly that cholera can be communicated through the water supply, Snow then extends his theory beyond its original area of applicability: is it possible that other infectious diseases are also transmitted in the same way? He considers four other epidemic diseases: yellow fever, malaria (intermittent fever, ague), dysentery, and typhoid fever. He was wrong about the first two and right about the second two. His reasoning in the cases where he

is wrong is interesting to quote, because he makes a plausible case:

> Yellow fever, which has been clearly proved by Dr. M'William and others to be a communicable disease, resembles cholera and the plague in flourishing best, as a general rule, on low alluvial soil, and also in spreading greatly where there is a want of personal cleanliness. This disease has more than once appeared in ships sailing up the river Plate, before they have had any communication with the shore. The most probable cause of this circumstance is, that the fresh water of this river, taken up from alongside the ship, contained the evacuations of patients with yellow fever in La Plata or other towns. ... (p. 127)
>
> Intermittent fevers are so fixed to particular places that they have deservedly obtained the name of endemics. They spread occasionally, however, much beyond their ordinary localities, and become epidemic. Intermittent fevers are undoubtedly often connected with a marshy state of the soil; for draining the land frequently causes their disappearance. They sometimes, however, exist as endemics, where there is no marshy land or stagnant water within scores of miles. Towards the end of the seventeenth century, intermittent fevers were, for the first time, attributed by Lancisi to noxious effluvia arising from marshes. These supposed effluvia, or marsh miasmata, as they were afterwards called, were thought to arise from decomposing vegetable and animal matter; but, as intermittent fevers have prevailed in many places where there was no decomposing vegetable or animal matter, this opinion has been given up in a great measure; still the belief in miasmata

or malaria[1] of some kind, as a cause of intermittents, is very general. It must be acknowledged, however, that there is no direct proof of the existence of malaria or miasmata, much less of their nature.

That preventive of ague, draining the land, must effect the water of a district quite as much as it affects the air, and there is direct evidence to prove that intermittent fever has, at all events in some cases, been caused by drinking the water of marshes, (pp. 129–130)

In the following paragraph, Snow, to explain the apparent absence of direct person-to-person contagion in malaria, makes an inspired guess: the malaria parasite, he speculates, must spend part of its life cycle outside the human body. Indeed it does, in the body of the *Anopheles* mosquito.

The communication of ague from person to person has not been observed, and supposing this disease to be communicable, it may be so only indirectly, for the materies morbi eliminated from one patient may require to undergo a process of development or procreation out of the body before it enters another patient, like certain flukes infesting some of the lower animals, and procreating by alternate generations. (p. 133)

Snow's explanation of why yellow fever breaks out on ships arriving on the River Plate before they even land is quite plausible, but is of course wrong. Similarly, the close identification of malaria with marshes had been known since the time of Hippocrates, and Snow's conjecture that it comes from drinking marsh water is also plausible but wrong. It is interesting that the Italian physician Lancisi (1654–1720), whose "effluvia" theory of malaria is quoted by Snow, also suggested that mosquitoes might spread malaria.[5] Snow does not refer to this idea, and we do not know what he thought of it.

---

[1] The word malaria as used by Snow is not the name of the disease but a term meaning "bad air."

What to Do? Measures to Prevent the Spread of Cholera

The last part of Snow's monograph gives his list of recommended measures for preventing the spread of cholera. His ideas did not win immediate acceptance from his medical contemporaries, who felt that he had made a good case for some influence of polluted water in cholera but continued to believe in "effluvia" theories as an alternate or contributing cause for a while. In any event, his recommendations on the water supply were adopted, and London was spared any further cholera epidemics.

The measures which are required for the prevention of cholera, and all diseases which are communicated in the same way as cholera, are of a very simple kind. They may be divided into those which may be carried out in the presence of an epidemic, and those which, as they require time, should be taken beforehand.

The measures which should be adopted during the presence of cholera may be enumerated as follows:

1st. The strictest cleanliness should be observed by those about the sick. There should be a hand-basin, water, and towel, in every room where there is a cholera patient, and care should be taken that they are frequently used by the nurse and other attendants, more particularly before touching any food.

2nd. The soiled bed linen and body linen of the patient should be immersed in water as soon as they are removed, until such time as they can be washed, lest the evacuations should become dry, and be wafted about as a fine dust. Articles of bedding and clothing which cannot be washed, should be exposed for some time to a temperature of 212° or upwards.

3rd. Care should be taken that the water employed for drinking and preparing food (whether it come from a pump-well, or be conveyed in pipes) is not contaminated with the contents of cesspools, house-drains, or sewers; or, in the event that water free from suspicion cannot be obtained, it should be well boiled, and if possible, also filtered. …

4th. When cholera prevails very much in the neighbourhood, all the provisions which are brought into the house should be well washed with clean water and exposed to a temperature of 212°F.; or at least they should undergo one of these processes, and be purified either by water or by fire. By being careful to wash the hands, and taking due precautions with regard to food, I consider that a person may spend his time amongst cholera patients without exposing himself to any danger.

5th. When a case of cholera or other communicable disease appears among persons living in a crowded room, the healthy should be removed to another apartment, where it is practicable, leaving only those who are useful to wait on the sick.

6th. As it would be impossible to clean out coal-pits, and establish privies and lavatories in them, or even to provide the means of eating a meal with anything like common decency, the time of working should be divided into periods of four hours instead of eight, so that the pitmen might go home to their meals, and be prevented from taking food into the mines.

7th. The communicability of cholera ought not to be disguised from the people, under the idea that the knowledge of it would cause a panic, or occasion the sick to be deserted.

The measures which can be taken beforehand to provide against cholera and other epidemic diseases, which are communicated in a similar way, are:

8th. To effect good and perfect drainage.

9th. To provide an ample supply of water quite free from contamination with the contents of sewers, cesspools, and house-drains, or the refuse of people who navigate the rivers.

10th. To provide model lodging-houses for the vagrant class, and sufficient house room for the poor generally. ...

11th. To inculcate habits of personal and domestic cleanliness among the people everywhere.

12th. Some attention should be undoubtedly directed to persons, and especially ships, arriving from infected places, in order to segregate the sick from the healthy. In the instance of cholera, the supervision would generally not require to be of long duration. ...

I feel confident, however, that by attending to the above-mentioned precautions, which I consider to be based on a correct knowledge of the cause of cholera, this disease may be rendered extremely rare, if indeed it may not be altogether banished from civilized countries. And the diminution of mortality ought not to stop with cholera. ... (pp. 133–137)

What Snow Overlooked

Snow's monograph ends with a paragraph stating that typhoid fever, which killed many more in England than did cholera, may also be controlled by the measures he proposed. This was right, and both diseases were soon brought under control.

We would like to close Snow's story with one more quotation, because it tells us something important about one aspect of scientific research. In order to make the problems we want to solve tractable, we need to limit the range of what we study. Yet by doing so we risk overlooking important possibilities not included within the narrowed scope of our inquiry.

Early in the monograph, Snow gives his reasons for believing that the "morbid matter" causing cholera reaches the digestive tract directly by ingestion, rather than through a preliminary systemic infection.

If any further proof were wanting than those above stated, that all the symptoms attending cholera, except those connected with the alimentary canal, depend simply on the physical alteration of the blood, and not on any cholera poison circulating in the system, it would only be necessary to allude to the effects of a weak saline solution injected into the veins in the stage of collapse. The shrunken skin becomes filled out, and loses its coldness and lividity; the countenance assumes a natural aspect; the patient is able to sit up, and for a time seems well. If the symptoms were caused by a poison circulating in the blood, and depressing the action of the heart, it is impossible that they should thus be suspended by an injection of warm water, holding a little carbonate of soda in solution. ... (p. 13)

Today it is recognized that cholera kills by dehydration and that if victims receive sufficient fluid either orally or intravenously the disease is rarely fatal. It is ironic that it should not have occurred to Snow that the observation he reported suggests a way of treating the disease. But Snow was not looking for a treatment of cholera, he was trying to establish how it is transmitted, and in that he succeeded.

From the Article About Snow's Work on Cholera

1. Find an example of deductive reasoning.

2. Find an example of inductive reasoning.

3. Find an example of an observation.

4. Find an example of a hypothesis or theory.

5. Find an example of an experiment.

6. Find an example of a predicted consequence.

7. Find an example of a verification of a predicted consequence.

Variables and Experimental Design

By John Oakes

Experimentation is the heart of science. Scientists conduct experiments for a variety of reasons. Sometimes they are attempting to optimize parameters for a desired result: for example, varying temperature, pressure, pH, and concentration of reactants in order to make a chemical reaction go with the greatest possible efficiency. At other times scientists do experiments as a kind of "fishing expedition," in order to look for possible cause-and-effect relationships. More rarely, scientists do trial-and-error experiments. The most common reason scientists conduct experiments is in order to test the validity of a hypothesis. In our formal scientific "method," this is the only reason for doing the experiments mentioned.

Some Definitions

By definition, a **scientific experiment** is a controlled measurement of the relationship between two variables. Two additional definitions of the word "experiment" will be helpful. An experiment is a controlled relational inquiry. An experiment is a contrived measurement of the effect of changing an independent variable on a dependent variable. In any experiment, there are three kinds of variables.

- **Independent Variable**: A hypothesis is a proposed cause-and-effect relationship. The proposed cause of the effect is the independent variable. As a rule, an experiment will have only one independent variable.
- **Dependent Variable**: In a "relational inquiry," the dependent variable is the effect. It is the thing affected by changing the independent variable. As a rule, an experiment will have only one dependent variable.
- **Control Variables**: In an experiment, a control variable is anything else that one can reasonably expect might affect the dependent variable, other than the independent variable. If all important variables are controlled, then any change in the dependent variable can be connected directly to the change in the independent variable, eliminating correlation as a possible explanation (see below).

In order to illustrate the three kinds of variables, consider the following scenario. (Never mind that this is a rather silly example.) Imagine for a moment that you are a guy. You have figured out what most guys eventually figure out, which is that women love to receive flowers. When a guy gives flowers to a girl, it produces happiness. Let us imagine further that you, as a guy, want to give flowers to your "significant other" in a way that produces the greatest possible amount of happiness. Whether you do this because you love her or in order to manipulate her for your own benefit is irrelevant to the illustration. In order to optimize the flower-giving event, you decide to use the scientific method. You will do a series of experiments to answer the question scientifically. If you are going to do a scientific experiment, then you first must identify your variables by type. First of all, the "level of happiness" is the effect of the flower-giving event. Therefore, it is the **dependent variable**. Let us propose a list of possible variables that might affect the level of happiness produced by giving flowers:

- color
- number of flowers
- species of flower
- timing/reason for giving flowers
- cost of the flowers
- location (at home, at work, etc.)

The thing about a scientific experiment is you can only vary one of the items on the list of possible causes at a time. If you vary both the color and the number of flowers, and the result is greater happiness, you will have learned literally nothing. You will not know what the cause was of the increased happiness was. Therefore, in order to begin your scientific investigation, you must choose one variable from the list. Let us say, for example, that you choose to experiment first with the relationship between color of flower and happiness. If that is the case, then according to the definition above, all other possible causes of happiness must be **control variables**, and the color of flower is the **independent variable**.

You will proceed as follows: In the first week, you will give six *white* roses, bought at a particular florist on Tuesday, and given in the afternoon at work. In the second week, you will give six *yellow* roses, bought

at the same florist on Tuesday, and given in the afternoon at work. In the third week, you will give six *red* roses, bought at the same florist, etc. … You continue testing for the effect of changing color for five weeks, collect your data, and conclude which is the optimal color of flower. Let us say it is red.

Now you choose a new independent variable/cause of happiness. This time, you will vary the type/specie of flower. In the second set of experiments, color will revert to being a control variable. The first week you will give six red *carnations*, bought at the same florist on Tuesday, and given in the afternoon at work. The second week you will give six red *roses*, bought at the same florist on Tuesday, and given in the afternoon at work, etc. … Of course, by now your significant other is probably on to the fact that something very fishy is going on.

A lighthearted example has been used here, but hopefully it has helped to make the definition and role of the three kinds of variables clear. If the entire procedure could be followed to its logical conclusion, assuming that your female friend has not figured out what is going on, you will have scientifically discovered the optimal set of flower-giving parameters to create the greatest happiness. With this much effort put in (never mind the money), you better stick with this girl for a long time.

A couple more things about dependent and independent variables deserve mentioning. For instance, even if this is the first time you have ever heard of the terms dependent and independent variables, and even if you feel a bit confused about the distinction, you probably have a better understanding of this concept than you think. First of all, whenever a graph is drawn, whether by a scientist or a nonscientist, the "y" axis is always the dependent variable, and the "x" axis is always the independent variable. Most of us have seen hundreds of graphs in our lives. If you can imagine a graph relating two variables, then the graph you imagine will have the dependent variable on the vertical axis, and the independent variable on the horizontal axis. Imagine a graph relating the two variables of cancer rate and exposure to alpha radiation. Probably your instinct says to put the cancer rate on the vertical axis and the exposure to alpha radiation on the horizontal axis. You have just discovered the dependent variable in this experiment.

Another way to decide on independent and dependent variables in a proposed experiment is to pose two statements to yourself. One of the two will be nonsense, while the other will reveal which is the dependent variable. "Cancer rate depends on radiation exposure" or "Radiation exposure depends on cancer rate." The second statement makes no sense. You have just discovered that cancer rate is the dependent variable. In addition, using the definition above, you now know that anything else that might affect cancer rate, other than exposure to radiation, is a control variable, and must be carefully controlled in order for you to reliably decide if radiation affects cancer rates.

Correlation Versus Cause and Effect

A second point can be made. One of the reasons scientists pay such strict attention to defining dependent, independent, and control variables is that we often find increasing one variable is found to be associated with an increase in another variable. Yet the relationship between the variables is one of correlation rather than cause and effect. **Correlation** occurs when two variables are statistically connected, but not as a direct cause and effect. For example, let us suppose that a health scientist discovers there is a greater incidence of heart disease for those living near high-voltage power lines (defined as living within 250 meters) than those living more than one kilometer from the same power lines. Does this mean that living near power lines causes heart disease? At first, this might seem to be the case, but it is very likely that this is correlation rather than cause and effect. If that is the case, then the statistical connection between the two would not

amount to an important scientific discovery. In fact, if one were to report the statistical connection, it might even produce a misleading interpretation.

Mathematical correlation is not the same as cause and effect. In science it is cause and effect, not correlation, that we look for. To illustrate the problem, it is not hard at all to imagine that those living near power lines have lower incomes, have much lower rates of good health insurance, and just possibly might have significantly poorer nutritional habits. Undoubtedly, this is the case in many communities for those living near power lines. If the scientist does not control her study for access to health care, diet, and possibly even income, then the experimental result is useless. It may be of great interest to the social scientist, but to the scientist, the result is of no value. Correlation counts for little—if anything—in science. The only thing that matters is cause and effect. The point to be made here is that careful attention to controlling all variables relevant to the result of a scientific experiment is the ideal way to assure that two variables are connected by cause and effect, and not merely by correlation.

Keys to Good Experimental Design

More than any other person, Galileo established the principles of scientific experimentation. He designed instruments with the specific intent of using them to measure variables. He set up contrived experimental setups, sliding objects down inclined planes, constructing the first thermometer, and more. An experiment is most commonly performed in a laboratory using instruments specifically designed for that purpose. Scientists use laboratories so they can create controlled environments to do their studies.

Below are a few key points to keep in mind when designing an experiment:

1. Variables should be as **specific and well defined** as possible.

2. **Quantifiable** variables are preferred over qualitative ones.
3. Careful attention is given to **controlling variables**, which can affect the dependent variable.
4. The experiment is designed to assure **reproducibility**. The experiment must be
5. performed enough times or on enough subjects to make the measured results reliable.

It takes some getting used to that scientists tend to study what appear to outsiders as very narrow—and therefore seemingly unimportant—questions. If you ask a scientist what he or she is studying, be prepared for the subject to be too narrowly defined to be interesting to you. The fact is, this is how science is done. It is not that scientists do not care about broader questions, but that they get at broader questions by investigating narrow ones. When designing an experiment, choosing carefully defined, narrow variables is key. Ensuring that the variables are quantifiable/measurable is not an absolute necessity, but it is always preferred.

For instance, let us imagine collecting wheat samples from a number of farm fields and measuring the protein concentration in each of the samples. Are we doing an experiment? The answer is no. There is only one variable—protein concentration—so by definition, this is not an experiment. A statistician may be interested in this data, but not a scientist. Science is not about the collection of facts. It is about cause-and-effect relationships between variables. Let us 'change our scenario slightly. Let us imagine collecting wheat samples from a number of farm fields and also doing a careful analysis of the mineral content in the different soils. Is this an experiment? Well, it seems so, because we have two variables: protein concentration and soil nutrients. Are our two variables "good" according to the parameters above? The answer is no, because one of our variables is not sufficiently narrow and well defined. Protein content is specific, well defined, and quantitative, but "soil nutrients" is none of these.

Now we'll improve the scenario. We need to choose a specific and well-defined aspect of soil nutrients as a possible cause of a change in protein content. For example, we might study wheat protein concentration versus soil nitrogen content. Now we have satisfied the first two criteria above for a good experiment. Nitrogen content is a specific, well-defined quantifiable variable.

However, as mentioned above, even if we can prove that soils with higher nitrogen content also have plants with a higher concentration of protein, this would not be a successful experiment. Without very careful attention to controlling other possible causes of increased wheat protein concentration, we learn nothing of use to science. Perhaps the same soils with higher nitrogen content also have higher content of organic materials, and this is the actual cause of the increased protein concentration in the wheat.

Let us now apply criterion #3. Probably we will have to perform a highly contrived experiment on a number of plots with carefully controlled soil. We will control for soil moisture, amount of sunlight, humidity, ambient temperature, and most important of all, the level of nutrients in the soil other than nitrogen, such as potassium and phosphorus. Other things to control are soil drainage, elevation above sea level, and more.

Last, in order to assure that our experiment produces usable results, care must be taken to assure that the experiment is reproducible. In order to assure this, many plants must be grown in each of a number of experiments. Perhaps hundreds of wheat plants each will be grown in plots, identical in every respect except for the nitrogen content. In order to make the experiment quantitative, plots with a close to zero nitrogen content, as well as with increasing amounts of nitrogen, will be planted. Data such as the following will be collected:

Nitrogen Content (mg/100g soil)	Protein Content
0	
10	
20	
30	
40	
50	

Methods of Variable Control

When designing an experiment, scientists want to control extraneous variables as carefully as possible. Depending on the circumstances, two kinds of variable control are used:

1. Total control.
2. Control "on average."

A variable has **total control** if it is caused to not vary at all. Of course, this is the ideal means of control. For example, if one were to study the effect of weight on the incidence of heart disease, clearly smoking, as a contributing factor toward heart disease, would have to be controlled for. An example of total control would be to eliminate all people who have ever smoked from the study. Doing so would eliminate any possible influence of this variable on the result. If one wanted to control for age, then all participants would be of the same age.

Sometimes total control is either impossible or too difficult to be practical. In this case, the scientist will attempt as well as possible to **control on average** for the variable. In other words, if there is a test and a control group, then the one designing the experiment will attempt to ensure that both groups, on average, have approximately the same condition with regard to the variable being controlled for.

As an example, let us return to our example of studying the effect of weight on heart disease rates. Medical scientists have sufficient experience to know that one factor in heart disease is genetics. There is no question that heredity plays a part in the incidence of heart disease. Of course, total control is always preferred, but the only way to have total control of genetic risk factors for heart disease is to use identical twins. This may not be impossible, but it is impractical and would probably make the study too expensive. How will the researcher control for genetics "on average?" He or she will have to choose a large enough study group so that, on average, genetic factors will be the same in both groups, genetics will average out. This will only work if it can be assumed that there is not a connected genetic cause, both of heart disease and of becoming overweight. We can see immediately that the requirement of controlling a variable on average is the use of a much greater number of people or test subjects in an experiment. It also raises the cost of the experiment.

Variables and Graphing

As has already been discussed, most experiments in a scientific context involve a dependent and an independent variable. In many cases, especially in the physical sciences such as physics and chemistry, the relationship between variables is expressed in the form of a mathematical relationship. The question to be addressed in this section is how the mathematical relationship between variables is arrived at. Although there are a number of ways of approaching this question, we will consider how graphing can be used to discover the mathematical form of the relationship between dependent and independent variables.

As examples of the mathematical relationship between cause and effect variables, consider some of the most famous equations in science. For example, there is Einstein's equation relating the energy content of matter to the amount of mass, which is given by the following equation:

$$E = mc^2.$$

According to this equation, the energy content of matter (E) is directly proportional to the amount of mass (m) and to the square of the speed of light (c^2). How could one arrive at this equation from experiment? This will be explained below. Worth noting here is that as a rule of thumb (not as a hard and fast rule!), scientists put the dependent variable in an experiment on the left side of the equal sign.

Another example of a very famous equation in science is that used to express the law of gravity, as discovered by Isaac Newton. The equation is as follows:

$$F = \frac{Gm_1 m_2}{d^2}.$$

To express this law in words, the force of attraction (F) between any two objects with mass is the dependent variable. Force is directly proportional to the mass of the two objects (m_1 and m_2) and inversely proportional to the square of the distance between the two objects (d^2). The letter G in this equation is the proportionality constant, which determines the size of the gravity force. G happens to have the value 6.67 x 10^{-11} Nm^2/kg^2. Two variables are **directly proportional** when increasing the independent variable causes the dependent variable to increase. Variables are **inversely proportional** when increasing the independent variable causes the dependent variable to decrease. In this case, a larger mass creates a larger force and a direct relationship, but a larger distance causes a smaller force due to gravity and an inverse relationship between the variables. Again, one might ask how one could actually use data on force and distance to discover the inverse square law relationship

between the variables. We will show how graphing of data can be used for this purpose.

Before doing this, we must discuss how scientists usually draw graphs. Although scientists do occasionally make bar graphs, pie graphs, and the like, the most common type of graph they use is the standard x, y graph. When two variables are graphed, one always plots **the dependent variable on the vertical axis** (traditionally known as the "y" axis) and **the independent variable on the horizontal axis** (traditionally known as the "x" axis).

For example, if a scientist were to study the average height of children as a function of age, she might gather the following data:

age (years)	height (inches)
0	22
1	31
2	34
3	38
4	40
5	43
6	46

Before making the graph, one must first ask which variable will be graphed on the vertical axis and which on the horizontal. Does the height of a child depend on age, or does the age of a child depend on height? Clearly, in this case, the height is the dependent variable. Therefore, the graph must be height (vertical, "y," dependent) versus age (horizontal, "x," independent).

The graph of this data is drawn below.

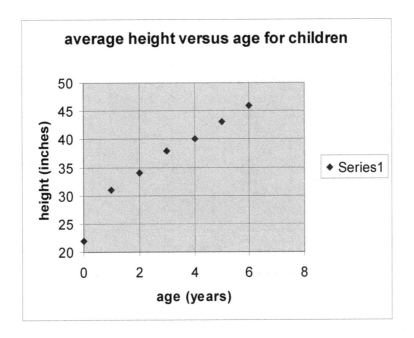

Once the data points are drawn, a smooth curve will be drawn through the data. If the data appear to be linear, in other words if the points appear to describe a straight line, a ruler will be used to draw a straight line, representing the best straight line fit to the data.

We will now address the issue of how scientists determine the mathematical relationship between two variables. One way to do this is to measure values of a dependent variable as a function of an independent variable, holding all the control variables constant, and discovering what graph of the variables produces a straight line. The principle used is as follows: **When two quantities are proportional, their graph is a straight line.**

For example, let us say that a scientist wanted to discover the relationship between the force of gravity between two masses and the distance between the two masses. In that case, one could measure data such as the following:

Force (Newtons)	Distance (meters)
0.0180	10
0.0080	15
0.0045	20
0.0029	25
0.0020	30
0.00113	40
0.00050	60

Notice that one can immediately detect an **inverse** relationship between the variables, because increasing distance is producing a decreasing force. Now, in order to discover the actual mathematical relationship between force and distance, we can make a series of graphs. The linear one will give us the correct relationship between the variables.

We can discover immediately that the relation between force of gravity and distance is that the force of gravity between two masses is proportional to the inverse of the distance squared. Of course, this is exactly what Isaac Newton discovered.

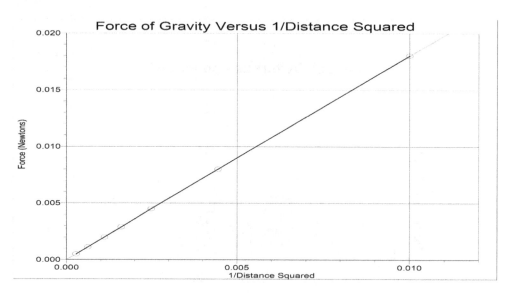

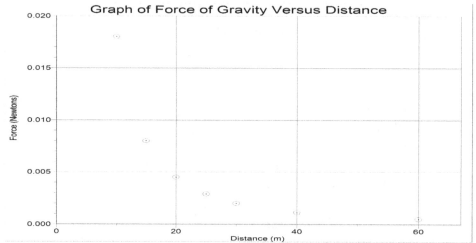

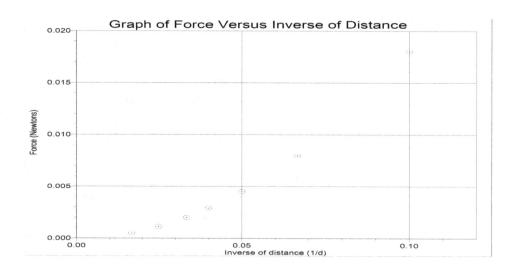

Section II
History of Science

Introduction

By John Oakes

You will begin your study of the history of science with a wonderful article by Carl Sagan. Sagan describes the Greek revolution in thinking as a transition from chaos to cosmos—from a view of the world with many competing gods to one of order and purpose. Here we get a glimpse of the greatest accomplishments in natural philosophy before the scientific revolution. As you read this article, ask yourself what it was about the Greek philosophers that allowed them to become, arguably, the greatest thinkers of the ancient world. What did their approach have in common with the modern scientific view of the world, and—perhaps more importantly—how was it different?

After establishing the greatest achievements in natural philosophy before the scientific revolution, you will next read a number of short biographies of the greatest figures in the scientific revolution. In doing so, you will be missing out on the vast majority of the history of science. How can a history of science not include the work of Lavoisier, Pasteur, Darwin, and Einstein? Presumably, you will be learning about these great scientists as part of this course, but in this section we are focusing in on how modern science came to be.

The article, titled *"The First Scientist,"* is a description of some of the proposed inventions of the visionary natural philosopher, Roger Bacon (1214–1292). Bacon is sometimes known as the first scientist, although it is more appropriate to describe him as the first philosopher of science. Bacon spent his entire fortune building instruments in order to do his crude "experiments" if that is what we can call his attempts to build ingenious machines. Having impoverished himself, he became a Franciscan monk, at least partially so that he could continue his experimental studies. Bacon probably invented gun powder. Actually, it would be more likely to say that he reinvented gunpowder, as the Chinese had already invented this explosive mixture. Bacon probably was unaware of this fact. Roger created the most accurate calendars of his time. All of this was achieved by Bacon 250 years before the scientific revolution. He did not do controlled experiments in the modern sense, but he strongly believed in the experimental approach to knowledge. It will be clear when you read the article that Bacon had not completely put behind him the superstitions of his day, despite his philosophy.

As you read about Copernicus, Kepler, Galileo, and Newton, ask yourself how they viewed the world. In what ways were they modern, and in what ways were they still medieval in their thinking? What might have been their unique contribution to the development of what we call science today?

Cosmos

By Carl Sagan

For thousands of years humans were oppressed—as some of us still are—by the notion that the universe is a marionette whose strings are pulled by a god or gods, unseen and inscrutable. Then, 2,500 years ago, there was a glorious awakening in Ionia: on Samos and the other nearby Greek colonies that grew up among the islands and inlets of the busy eastern Aegean Sea.[1] Suddenly there were people who believed that everything was made of atoms; that human beings and other animals had sprung from simpler forms; that diseases were not caused by demons or the gods; that the Earth was only a planet going around the Sun. And that the stars were very far away.

This revolution made Cosmos out of Chaos. The early Greeks had believed that the first being was Chaos, corresponding to the phrase in Genesis in the same context, "without form." Chaos created and then mated with a goddess called Night, and their offspring eventually produced all the gods and men. A universe created from Chaos was in perfect keeping with the Greek belief in an unpredictable nature run by capricious gods. But in the sixth century B.C., in Ionia, a new concept developed, one of the great ideas of the human species. The universe is knowable, the ancient Ionians argued, because it exhibits an internal order: there are regularities in nature that permit its secrets to be uncovered. Nature is not entirely unpredictable; there are rules even she must obey. This ordered and admirable character of the universe was called Cosmos.

But why Ionia, why in these unassuming and pastoral landscapes, these remote islands and inlets of the Eastern Mediterranean? Why not in the great cities of India or Egypt, Babylonia, China or Mesoamerica? China had an astronomical tradition millennia old; it invented paper and printing, rockets, clocks, silk, porcelain, and ocean-going navies. Some historians argue it was nevertheless too traditionalist a society, too unwilling to adopt innovations. Why not India, an extremely rich, mathematically gifted culture? Because, some historians maintain, of a rigid fascination with the idea of an infinitely old universe condemned to an endless cycle of deaths and rebirths, of souls and universes, in which nothing fundamentally new could ever happen. Why not Mayan and Aztec societies, which were accomplished in astronomy and captivated, as the Indians were, by large numbers? Because, some historians declare, they lacked the aptitude or impetus for mechanical invention. The

---

[1] As an aid to confusion, Ionia is not in the Ionian Sea; it was named by colonists from the coast of the Ionian Sea.

Mayans and the Aztecs did not even—except for children's toys—invent the wheel.

The Ionians had several advantages. Ionia is an island realm. Isolation, even if incomplete, breeds diversity. With many different islands, there was a variety of political systems. No single concentration of power could enforce social and intellectual conformity in all the islands. Free inquiry became possible. The promotion of superstition was not considered a political necessity. Unlike many other cultures, the Ionians were at the crossroads of civilizations, not at one of the centers. In Ionia, the Phoenician alphabet was first adapted to Greek usage and widespread literacy became possible. Writing was no longer a monopoly of the priests and scribes. The thoughts of many were available for consideration and debate. Political power was in the hands of the merchants, who actively promoted the technology on which their prosperity depended. It was in the Eastern Mediterranean that African, Asian, and European civilizations, including the great cultures of Egypt and Mesopotamia, met and cross-fertilized in a vigorous and heady confrontation of prejudices, languages, ideas and gods. What do you do when you are faced with several different gods each claiming the same territory? The Babylonian Marduk and the Greek Zeus was each considered master of the sky and king of the gods. You might-decide that Marduk and Zeus were really the same. You might also decide, since they had quite different attributes, that one of them was merely invented by the priests. But if one, why not both?

And so it was that the great idea arose, the realization that there might be a way to know the world without the god hypothesis; that there might be principles, forces, laws of nature through which the world could be understood without attributing the fall of every sparrow to the direct intervention of Zeus

China and India and Mesoamerica would, I think, have tumbled to science too, if only they had been given a little more time. Cultures do not develop with identical rhythms or evolve in lockstep. They arise at different times and progress at different rates. The scientific world view works so well, explains so much and resonates so harmoniously with the most advanced parts of our brains that in time, I think, virtually every culture on the Earth, left to its own devices, would have discovered science. Some culture had to be first. As it turned out, Ionia was the place where science was born.

Between 600 and 400 B.C., this great revolution in human thought began. The key to the revolution was the hand. Some of the brilliant Ionian thinkers were the sons of sailors and farmers and weavers. They were accustomed to poking and fixing, unlike the priests and scribes of other nations, who, raised in luxury, were reluctant to dirty their hands. They rejected superstition, and they worked wonders. In many cases we have only fragmentary or secondhand accounts of what happened. The metaphors used then may be obscure to us now. There was almost certainly a conscious effort a few centuries later to suppress the new insights. The leading figures in this revolution were men with Greek names, largely unfamiliar to us today, but the truest pioneers in the development of our civilization and our humanity.

The first Ionian scientist was Thales of Miletus, a city in Asia across a narrow channel of water from the island of Samos. He had traveled in Egypt and was conversant with the knowledge of Babylon. It is said that he predicted a solar eclipse. He learned how to measure the height of a pyramid from the length of its shadow and the angle of the Sun above the horizon, a method employed today to determine the heights of the mountains of the Moon. He was the first to prove geometric theorems of the sort codified by Euclid three centuries later-for example, the proposition that the angles at the base of an isosceles triangle are equal. There is a clear continuity of intellectual effort from Thales to Euclid to Isaac Newton's purchase of the Elements *of Geometry* at Stourbridge Fair in 1663 (p. 68), the event that precipitated modern science and technology.

Thales attempted to understand the world without invoking the intervention of the gods. Like the Babylonians, he believed the world to have once been water. To explain the dry land, the Babylonians added that Marduk had placed a mat on the face of the waters and piled dirt upon it.[2] Thales held a similar view, but, as Benjamin Farrington said, "left Marduk out." Yes, everything was once water, but the Earth formed out of the oceans by a natural process—similar, he thought, to the silting he had observed at the delta of the Nile. Indeed, he thought that water was a common principle underlying all of matter, just as today we might say the same of electrons, protons and neutrons, or of quarks. Whether Thales' conclusion was correct is not as important as his approach: The world was not made by the gods, but instead was the work of material forces interacting in Nature. Thales brought back from Babylon and Egypt the seeds of the new sciences of astronomy and geometry, sciences that would sprout and grow in the fertile soil of Ionia.

Very little is known about the personal life of Thales, but one revealing anecdote is told by Aristotle in his Politics:

[Thales] was reproached for his poverty, which was supposed to show that philosophy is of no use. According to the story, he knew by his skill [in interpreting the heavens] while it was yet winter that there would be a great harvest of olives in the coming year; so, having a little money, he gave deposits for the use of all the olive-presses in Chios and Miletus, which he hired at a low price because no one bid against him. When the harvest time came, and many were wanted all at once, he let them out at any rate which he pleased and made a quantity of money. Thus he showed the world philosophers can easily be rich if they like, but that their ambition is of another sort.

He was also famous as a political sage, successfully urging the Milesians to resist assimilation by Croesus, King of Lydia, and unsuccessfully urging a federation of all the island states of Ionia to oppose the Lydians.

Anaximander of Miletus was a friend and colleague of Thales, one of the first people we know of to do an experiment. By examining the moving shadow cast by a vertical stick he determined accurately the length of the year and the seasons. For ages men had used sticks to club and spear one another. Anaximander used one to measure time. He was the first person in Greece to make a sundial, a map of the known world and a celestial globe that showed the patterns of the constellations. He believed the Sun, the Moon and the stars to be made of fire seen through moving holes in the dome of the sky, probably a much older idea. He held the remarkable view that the Earth is not suspended or supported from the heavens, but that it remains by itself at the center of the universe; since it was equidistant from all places on the "celestial sphere," there was no force that could move it.

He argued that we are so helpless at birth that, if the first human infants had been put into the world on their own, they would immediately have died. From this Anaximander concluded that human beings arose

---

[2] There is some evidence that the antecedent, early Sumerian creation myths were largely naturalistic explanations, later codified around 1000 B.C. in the Enuma *elish* ("When on high," the first words of the poem); but by then the gods had replaced Nature, and the myth offers a theogony, not a cosmogony. The *Enuma elish* is reminiscent of the Japanese and Ainu myths in which an originally muddy cosmos is beaten by the wings of a bird, separating the land from the water. A Fijian creation myth says: "Rokomautu created the land. He scooped it up out of the bottom of the ocean in great handfuls and accumulated it in piles here and there. These are the Fiji Islands." The distillation of land from water is a natural enough idea for island and seafaring peoples.

from other animals with more self-reliant newborns: He proposed the spontaneous origin of life in mud, the first animals being fish covered with spines. Some descendants of these fishes eventually abandoned the water and moved to dry land, where they evolved into other animals by the transmutation of one form into another. He believed in an infinite number of worlds, all inhabited, and all subject to cycles of dissolution and regeneration. "Nor," as Saint Augustine ruefully complained, "did he, any more than Thales, attribute the cause of all this ceaseless activity to a divine mind."

In the year 540 B.C. or thereabouts, on the island of Samos, there came to power a tyrant named Polycrates. He seems to have started as a caterer and then gone on to international piracy. Polycrates was a generous patron of the arts, sciences and engineering. But he oppressed his own people; he made war on his neighbors; he quite rightly feared invasion. So he surrounded his capital city with a massive wall, about six kilometers long, whose remains stand to this day. To carry water from a distant spring through the fortifications, he ordered a great tunnel built. A kilometer long, it pierces a mountain. Two cuttings were dug from either end which met almost perfectly in the middle. The project took about fifteen years to complete, a testament to the civil engineering of the day and an indication of the extraordinary practical capability of the Ionians. But there is another and more ominous side to the enterprise: it was built in part by slaves in chains, many captured by the pirate ships of Polycrates.

This was the time of Theodorus, the master engineer of the age, credited among the Greeks with the invention of the key, the ruler, the carpenters square, the level, the lathe, bronze casting and central heating. Why are there no monuments to this man? Those who dreamed and speculated about the laws of Nature talked with the technologists and the engineers. They were often the same people. The theoretical and the practical were one.

About the same time, on the nearby island of Cos, Hippocrates was establishing his famous medical tradition, now barely remembered because of the Hippocratic oath. It was a practical and effective school of medicine, which Hippocrates insisted had to be based on the contemporary equivalent of physics and chemistry.[3] But it also had its theoretical side. In his book On *Ancient Medicine,* Hippocrates wrote: "Men think epilepsy divine, merely because they do not understand it. But if they called everything divine which they do not understand, why, there would be no end of divine things."

In time, the Ionian influence and the experimental method spread to the mainland of Greece, to Italy, to Sicily. There was once a time when hardly anyone believed in air. They knew about breathing, of course, and they thought the wind was the breath of the gods. But the idea of air as a static, material but invisible substance was unimagined. The first recorded experiment on air was performed by a physician[4] named Empedocles, who flourished around 450 B.C. Some accounts claim he identified himself as a god. But perhaps it was only that he was so clever that others thought him a god. He believed that light travels very fast, but not infinitely fast. He taught that there was once a much greater variety of living things on the Earth, but that many races of beings "must have been unable to beget and continue their kind. For in the case of every species that exists, either craft or courage or speed has from the beginning of its existence protected and preserved it." In this attempt to explain the

---

[3] And astrology, which was then widely regarded as a science. In a typical passage, Hippocrates writes: "One must also guard against the risings of the stars, especially of the Dog Star [Sirius], then of Arcturus, and also of the setting of the Pleiades."

[4] The experiment was performed in support of a totally erroneous theory of the circulation of the blood, but the idea of performing any experiment to probe Nature is the important innovation.

lovely adaptation of organisms to their environments, Empedocles, like Anaximander and Democritus (see below), clearly anticipated some aspects of Darwin's great idea of evolution by natural selection.

Empedocles performed his experiment with a household implement people had used for centuries, the so-called *clepsydra* or "water thief," which was used as a kitchen ladle. A brazen sphere with an open neck and small holes in the bottom, it is filled by immersing it in water. If you pull it out with the neck uncovered, the water pours out of the holes, making a little shower. But if you pull it out properly, with your thumb covering the neck, the water is retained within the sphere until you lift your thumb. If you try to fill it with the neck covered, nothing happens. Some material substance must be in the way of the water. We cannot see such a substance. What could it be? Empedocles argued that it could only be air. A thing we cannot see can exert pressure, can frustrate my wish to fill a vessel with water if I were dumb enough to leave my finger on the neck. Empedocles had discovered the invisible. Air, he thought, must be matter in a form so finely divided that it could not be seen.

Empedocles is said to have died in an apotheotic fit by leaping into the hot lava at the summit caldera of the great volcano of Aetna. But I sometimes imagine that he merely slipped during a courageous and pioneering venture in observational geophysics.

This hint, this whiff, of the existence of atoms was carried much further by a man named Democritus, who came from the Ionian colony of Abdera in northern Greece. Abdera was a kind of joke town. If in 430 B.C. you told a story about someone from Abdera, you were guaranteed a laugh. It was in a way the Brooklyn of its time. For Democritus all of life was to be enjoyed and understood; understanding and enjoyment were the same thing. He said that "a life without festivity is a long road without an inn." Democritus may have come from Abdera, but he was no dummy. He believed that a large number of worlds had formed spontaneously out of diffuse matter in space, evolved and then decayed.

At a time when no one knew about impact craters, Democritus thought that worlds on occasion collide; he believed that some worlds wandered alone through the darkness of space, while others were accompanied by several suns and moons; that some worlds were inhabited, while others had no plants or animals or even water; that the simplest forms of life arose from a kind of primeval ooze. He taught that perception—the reason, say, I think there is a peri in my hand—was a purely physical and mechanistic process; that thinking and feeling were attributes of matter put together in a sufficiently fine and complex way and not due to some spirit infused into matter by the gods.

Democritus invented the word atom, Greek for "unable to be cut." Atoms were the ultimate particles, forever frustrating our attempts to break them into smaller pieces. Everything, he said, is a collection of atoms, intricately assembled. Even we "Nothing exists," he said, "but atoms and the void."

When we cut an apple, the knife must pass through empty spaces between the atoms, Democritus argued. If there were no such empty spaces, no void, the knife would encounter the impenetrable atoms, and the apple could not be cut. Having cut a slice from a cone, say, let us compare the cross sections of the two pieces. Are the exposed areas equal? No, said Democritus. The slope of the cone forces one side of the slice to have a slightly smaller cross section than the other. If the two areas were exactly equal, we would have a cylinder, not a cone. No matter how sharp the knife, the two pieces have unequal cross sections. Why? Because, on the scale of the very small, matter exhibits some irreducible roughness. This fine scale of roughness Democritus identified with the world of the atoms. His arguments were not those we use today, but they were subtle and elegant, derived from everyday life. And his conclusions were fundamentally correct.

In a related exercise, Democritus imagined calculating the volume of a cone or a pyramid by a very large number of extremely small stacked plates tapering in size from the base to the apex. He had stated the problem that, in mathematics, is called the theory of limits. He was knocking at the door of the differential and integral calculus, that fundamental tool for understanding the world that was not, so far as we know from written records, in fact discovered until the time of Isaac Newton. Perhaps if Democritus' work had not been almost completely destroyed, there would have been calculus by the time of Christ.[5]

Thomas Wright marveled in 1750 that Democritus had believed the Milky Way to be composed mainly of unresolved stars: "long before astonomy reaped any benefit from the improved sciences of optics; [he] saw, as we may say, through the eye of reason, full as far into infinity as the most able astronomers in more advantageous times have done since." Beyond the Milk of Hera, past the Backbone of Night, the mind of Democritus soared.

As a person, Democritus seems to have been somewhat unusual. Women, children and sex discomfited him, in part because they took time away from thinking. But he valued friendship, held cheerfulness to be the goal of life and devoted a major philosophical inquiry to the origin and nature of enthusiasm. He journeyed to Athens to visit Socrates and then found himself too shy to introduce himself. He was a close friend of Hippocrates. He was awed by the beauty and elegance of the physical world. He felt that poverty in a democracy was preferable to wealth in a tyranny. He believed that the prevailing religions of his time were evil and that neither immortal souls nor immortal gods exist: "Nothing exists, but atoms and the void."

There is no record of Democritus having been persecuted for his opinions—but then, he came from Abdera. However, in his time the brief tradition of tolerance for unconventional views began to erode and then to shatter. People came to be punished for having unusual ideas. A portrait of Democritus is now on the Greek hundred-drachma bill. But his insights

were suppressed, his influence on history made minor. The mystics were beginning to win.

Anaxagoras was an Ionian experimentalist who flourished around 450 B.C. and lived in Athens. He was a rich man, indifferent to his wealth but passionate about science. Asked what was the purpose of life, he replied, "the investigation of the Sun, the Moon, and the heavens," the reply of a true astronomer. He performed a clever experiment in which a single drop of white liquid, like cream, was shown not to lighten perceptibly the contents of a great pitcher of dark liquid, like wine. There must, he concluded, be changes deducible by experiment that are too subtle to be perceived directly by the senses.

Anaxagoras was not nearly so radical as Democritus. Both were thoroughgoing materialists, not in prizing possessions but in holding that matter alone provided the underpinnings of the world. Anaxagoras believed in a special mind substance and disbelieved in the existence of atoms. He thought humans were more intelligent than other animals because of our hands, a very Ionian idea.

He was the first person to state clearly that the Moon shines by reflected light, and he accordingly devised a theory of the phases of the Moon. This doctrine was so dangerous that the manuscript describing it had to be circulated in secret, an Athenian *samizdat*. It was not in keeping with the prejudices of the time to explain the phases or eclipses of the Moon by the relative geometry of the Earth, the Moon and the self-luminous Sun. Aristotle, two generations later, was content to argue that those things happened because it was the nature of the Moon to have phases and eclipses—mere verbal juggling, an explanation that explains nothing.

The prevailing belief was that the Sun and Moon were gods. Anaxagoras held that the Sun and stars are fiery stones. We do not feel the heat of the stars because they are too far away. He also thought that the Moon has mountains (right) and inhabitants (wrong). He held that the Sun was so huge that it was probably larger than the Peloponnesus, roughly the southern

---

[5] The frontiers of the calculus were also later breached by Eudoxus and Archimedes.

third of Greece. His critics thought this estimate excessive and absurd.

Anaxagoras was brought to Athens by Pericles, its leader in its rime of greatest glory, but also the man whose actions led to the Peloponnesian War, which destroyed Athenian democracy. Pericles delighted in philosophy and science, and Anaxagoras was one of his principal confidants. There are those who think that in this role Anaxagoras contributed significantly to the greatness of Athens. But Pericles had political problems. He was too powerful to be attacked directly, so his enemies attacked those close to him. Anaxagoras was convicted and imprisoned for the religious crime of impiety—because he had taught that the Moon was made of ordinary matter, that it was a place, and that the Sun was a red-hot stone in the sky. Bishop John Wilkins commented in 1638 on these Athenians: "Those zealous idola-tors [counted] it a great blasphemy to make their God a stone, whereas notwithstanding they were so senseless in their adoration of idols as to make a stone their God." Pericles seems to have engineered Anaxagoras' release from prison, but it was too late. In Greece the tide was turning, although the Ionian tradition continued in Alexandrian Egypt two hundred years later.

The great scientists from Thales to Democritus and Anaxagoras have usually been described in history or philosophy books as "Presocrarics," as if their main function was to hold the philosophical fort until the advent of Socrates, Plato, and Aristotle and perhaps influence them a little. Instead, the old Ionians represent a different and largely contradictory tradition, one in much better accord with modern science. That their influence was felt powerfully for only two or three centuries is an irreparable loss for all those human beings who lived between the Ionian Awakening and the Italian Renaissance.

Perhaps the most influential person ever associated with Samos was Pythagoras,[6] a contemporary of Polycrates in the sixth century B.C. According to local tradition, he lived for a time in a cave on the Samian Mount Kerkis, and was the first person in the history of the world to deduce that the Earth is a sphere. Perhaps he argued by analogy with the Moon and the Sun, or noticed the curved shadow of the Earth on the Moon during a lunar eclipse, or recognized that when ships leave Samos and recede over the horizon, their masts disappear last.

He or his disciples discovered the Pythagorean theorem: the sum of the squares of the shorter sides of a right triangle equals the square of the longer side. Pythagoras did not simply enumerate examples of this theorem; he developed a method of mathematical deduction to prove the thing generally. The modern tradition of mathematical argument, essential to all of science, owes much to Pythagoras. It was he who first used the word *Cosmos* to denote a well-ordered and harmonious universe, a world amenable to human understanding.

Many Ionians believed the underlying harmony of the universe to be accessible through observation and experiment, the method that dominates science today. However, Pythagoras employed a very different method. He taught that the laws of Nature could be deduced by pure thought. He and his followers were not fundamentally experimentalists.[7] They

---

[6] The sixth century B.C. was a time of remarkable intellectual and spiritual ferment across the planet. Not only was it the time of Thales, Anaximander, Pythagoras and others in Ionia, but also the time of the Egyptian Pharaoh Necho who caused Africa to be circumnavigated, of Zoroaster in Persia, Confucius and Lao-tse in China, the Jewish prophets in Israel, Egypt and Babylon, and Gautama Buddha in India. It is hard to think these activities altogether unrelated.

[7] Although there were a few welcome exceptions. The Pythagorean fascination with whole number rations in musical harmonies seems clearly to be based on observation, or

were mathematicians. And they were thoroughgoing mystics. According to Berrrand Russell, in a perhaps uncharitable passage, Pythagoras "founded a religion, of which the main tenets were the transmigration of souls and the sinfulness of eating beans. His religion was embodied in a religious order, which, here and there, acquired control of the State and established a rule of the saints. But the unregenerate hankered after beans, and sooner or later rebelled."

The Pythagoreans delighted in the certainty of mathematical demonstration, the sense of a pure and unsullied world accessible to the human intellect, a Cosmos in which the sides of right triangles perfectly obey simple mathematical relationships. It was in striking contrast to the messy reality of the workaday world. They believed that in their mathematics they had glimpsed a perfect reality, a realm of the gods, of which our familiar world is but an imperfect reflection. In Plato's famous parable of the cave, prisoners were imagined tied in such a way that they saw only the shadows of passersby and believed the shadows to be real—never guessing the complex reality that was accessible if they would but turn their heads. The Pythagoreans would powerfully influence Plato and, later, Christianity.

They did not advocate the free confrontation of conflicting points of view. Instead, like all orthodox religions, they practiced a rigidity that prevented them from correcting their errors. Cicero wrote:

In discussion it is not so much weight of authority as force of argument that should be demanded. Indeed, the authority of those who profess to teach is often a positive hindrance to those who desire to learn; they cease to employ their own judgment, and take what they perceive to be the verdict of their chosen master as settling the question. In fact I am not disposed to approve the practice traditionally ascribed to the Pythagoreans, who, when questioned as to the grounds of any assertion that they advanced in debate, are said to have been accustomed to reply "The Master said so," "the Master" being Pythagoras. So potent was an opinion already decided, making authority prevail unsupported by reason.

The Pythagoreans were fascinated by the regular solids, symmetrical three-dimensional objects all of whose sides are the same regular polygon. The cube is the simplest example, having six squares as sides. There are an infinite number of regular polygons, but only five regular solids. (The proof of this statement, a famous example of mathematical reasoning, is given in Appendix 2.) For some reason, knowledge of a solid called the dodecahedron having twelve pentagons as sides seemed to them dangerous. It was mystically associated with the Cosmos. The other four regular solids were identified, somehow, with the four "elements" then imagined to constitute the world: earth,- fire, air and water. The fifth regular solid must then, they thought, correspond to some fifth element that could only be the substance of the heavenly bodies. (This notion of a fifth essence is the origin of our word *quintessence.)* Ordinary people were to be kept ignorant of the dodecahedron.

In love with whole numbers, the Pythagoreans believed all things could be derived from them, certainly all other numbers. A crisis in doctrine arose when they discovered that the square root of two (the ratio of the

---

even experiment on the sounds issued from plucked strings. Empedocles was, at least in part, a Pythagorean. One of Pythagoras' students, Alcmaeon, is the first person known to have dissected a human body; he distinguished between arteries and veins, was the first to discover the optic nerve and the eustachain tubes, and identified the brain as the seat of the intellect (a contention later denied by Aristotle, who placed intelligence in the heart, and then revived by Herophilus of Chalcedon). He also founded the science of embryology. But Alcmaeon's zest for the impure was not shared by most of his Pythagorean colleagues in later times.

diagonal to the side of a square) was irrational, that 2 cannot be expressed accurately as the ratio of any two whole numbers, no matter how big these numbers are. Ironically this discovery (reproduced in Appendix 1) was made with the Pythagorean theorem as a tool. "Irrational" originally meant only that a number could not be expressed as a ratio. But for the Pythagoreans it came to mean something threatening, a hint that their world view might not make sense, which is today the other meaning of "irrational." Instead of sharing these important mathematical discoveries, the Pythagoreans suppressed knowledge of √2 and the dodecahedron. The outside world was not to know.[8] Even today there are scientists opposed to the popularization of science: the sacred knowledge is to be kept within the cult, unsullied by public understanding.

The Pythagoreans believed the sphere to be "perfect," all points on its surface being at the same distance from its center. Circles were also perfect. And the Pythagoreans insisted that planets moved in circular paths at constant speeds. They seemed to believe that moving slower or faster at different places in the orbit would be unseemly; noncircular motion was somehow flawed, unsuitable for the planets, which, being free of the Earth, were also deemed "perfect."

The pros and cons of the Pythagorean tradition can be seen clearly in the life's work of Johannes Kepler (Chapter 3). The Pythagorean idea of a perfect and mystical world, unseen by the senses, was readily accepted by the early Christians and was an integral component of Kepler's early training. On the one hand, Kepler was convinced that mathematical harmonies exist in nature (he wrote that "the universe was stamped with the adornment of harmonic proportions"); that simple numerical relationships must

determine the motion of the planets. On the other hand, again following the Pythagoreans, he long believed that only uniform circular motion was admissible. He repeatedly found that the observed planetary motions could not be explained in this way, and repeatedly tried again. But unlike many Pythagoreans, he believed in observations and experiment in the real world. Eventually the detailed observations of the apparent motion of the planets forced him to abandon the idea of circular paths and to realize that planets travel in ellipses. Kepler was both inspired in his search for the harmony of planetary motion and delayed for more than a decade by the attractions of Pythagorean doctrine.

A disdain for the practical swept the ancient world. Plato urged astronomers to think about the heavens, but not to waste their time observing them. Aristotle believed that: "The lower sort are by nature slaves, and it is better for them as for all inferiors that they should be under the rule of a master. … The slave shares in his master's life; the artisan is less closely connected with him, and only attains excellence in proportion as he becomes a slave. The meaner son of mechanic has a special and separate slavery." Plutarch wrote: "It does not of necessity follow that, if the work delight you with its grace, the one who wrought it is worthy of esteem." Xenophon's opinion was: "What are called the mechanical arts carry a social stigma and are rightly dishonoured in our cities." As a result of such attitudes, the brilliant and promising Ionian experimental method was largely abandoned for two thousand years. Without experiment, there is no way to choose among contending hypotheses, no way for science to advance. The antiempirical taint of the Pythagoreans survives to this day. But why? Where did this distaste for experiment come from?

An explanation for the decline of ancient science has been put forward by the historian of science, Benjamin Farrington: The mercantile tradition, which led to Ionian science, also led to a slave economy. The owning of slaves was the road to wealth and power.

---

[8] A Pythagorean named Hippasus published the secret of the "sphere with twelve pentagons, "the dodecahedron. When he later died in a shipwreck, we are told, his fellow Pythagoreans remarked on the justice of the punishment. His book has not survived.

Polycrates' fortifications were built by slaves. Athens in the rime of Pericles, Plato and Aristotle had a vast slave population. All the brave Athenian talk about democracy applied only to a privileged few. What slaves characteristically perform is manual labor. But scientific experimentation is manual labor, from which the slaveholders are preferentially distanced; while it is only the slaveholders—politely called "gentle-men" in some societies—who have the leisure to do science. Accordingly, almost no one did science. The Ionians were perfectly able to make machines of some elegance. But the availability of slaves undermined the economic motive for the development of technology. Thus the mercantile tradition contributed to the great Ionian awakening around 600 B.C., and, through slavery, may have been the cause of its decline some two centuries later. There are great ironies here.

Similar trends are apparent throughout the world. The high point in indigenous Chinese astronomy occurred around 1280, with the work of Kuo Shou-ching, who used an observational baseline of 1,500 years and improved both astronomical instruments and mathematical techniques for computation. It is generally thought that Chinese astronomy thereafter underwent a steep decline. Nathan Sivin believes that the reason lies at least partly "in increasing rigidity of elite attitudes, so that the educated were less inclined to be curious about techniques and less willing to value science as an appropriate pursuit for a gentleman." The occupation of astronomer became a hereditary office, a practice inconsistent with the advance of the subject. Additionally, "the responsibility for the evolution of astronomy remained centered in the Imperial Court and was largely abandoned to foreign technicians," chiefly the Jesuits, who had introduced Euclid and Copernicus to the astonished Chinese, but who, after the censorship of the latter's book, had a vested interest in disguising and suppressing heliocentric cosmology. Perhaps science was stillborn in Indian, Mayan and Aztec civilizations for the same reason it declined in Ionia, the pervasiveness of the slave economy. A major

problem in the contemporary (political) Third World is that the educated classes tend to be the children of the wealthy, with a vested interest in the status quo, and are unaccustomed either to working with their hands or to challenging conventional wisdom. Science has been very slow to take root.

Plato and Aristotle were comfortable in a slave society. They offered justifications for oppression. They served tyrants. They taught the alienation of the body from the mind (a natural enough ideal in a slave society); they separated matter from thought; they divorced the Earth from the heavens—divisions that were to dominate Western thinking for more than twenty centuries. Plato, who believed that "all things are full of gods "actually used the metaphor of slavery to connect his politics with his cosmology. He is said to have urged the burning of all the books of Democritus (he had a similar recommendation for the books of Homer), perhaps because Democritus did not acknowledge immortal souls or immortal gods or Pythagorean mysticism or because he believed in an infinite number of worlds. Of the seventy-three books Democritus is said to have written, covering all of human knowledge, not a single work survives. All we know is from fragments, chiefly on ethics, and secondhand accounts. The same is true of almost all the other ancient Ionian scientists.

In the recognition by Pythagoras and Plato that the Cosmos is knowable, that there is a mathematical underpinning to nature, they greatly advanced the cause of science. But in the suppression of disquieting facts, the sense that science should be kept for a small elite, the distaste for experiment, the embrace of mysticism and the easy acceptance of slave societies, they set back the human enterprise. After a long mystical sleep in which the tools of scientific inquiry lay moldering, the Ionian approach, in some cases transmitted through scholars at the Alexandrian Library, was finally rediscovered. The Western world reawakened. Experiment and open inquiry became once more respectable. Forgotten books and fragments were again read. Leonardo and Columbus and Copernicus were inspired by or independently retraced parts of this ancient Greek tradition. There is in our time much Ionian science, although not in politics and religion, and a fair amount of courageous free inquiry. But there are also appalling superstitions and deadly ethical ambiguities. We are flawed by ancient contradictions.

The Platonists and their Christian successors held the peculiar notion that the Earth was tainted and somehow nasty, while the heavens were perfect and divine. The fundamental idea that the Earth is a planet, that we are citizens of the Universe, was rejected and forgotten. This idea was first argued by Aristarchus, born on Samos three centuries after Pythagoras. Aristarchus was one of the last of the Ionian scientists. By this time, the center or intellectual enlightenment had moved to the great Library of Alexandria. Aristarchus was the first person to hold that the Sun rather than the Earth is at the center of the planetary system, that all the planets go around the Sun rather than the Earth. Typically, his writings on this matter are lost. From the size of the Earths shadow on the Moon during a lunar eclipse, he deduced that the Sun had to be much larger than the Earth, as well as very far away. He may then have reasoned that it is absurd for so large a body as the Sun to revolve around so small a body as the Earth. He put the Sun at the center, made the Earth rotate on its axis once a day and orbit the Sun once a year.

It is the same idea we associate with the name of Copernicus, whom Galileo described as the "restorer and confirmer," not the inventor, of the heliocentric hypothesis.[9] For most of the 1,800 years between Aristarchus and Copernicus nobody knew the correct disposition of the planets, even though it had been laid out perfectly clearly around 280 B.C. The idea outraged some of Aristarchus' contemporaries. There were cries, like those voiced about Anaxagoras and Bruno and Galileo, that he be condemned for impiety. The resistance to Aristarchus and Copernicus, a kind of geocentrism in everyday life, remains with us: we still talk about the Sun "rising" and the Sun "setting." It is 2,200 years since Aristarchus, and our language still pretends that the Earth does not turn.

---

9 Copernicus may have gotten the idea from reading about Aristotle. Recently discovered classical texts were a source of great excitement Italian universities when Copernicus went to medical, school that the manuscript of his book, Copernicus mentioned Aristarchus but he omitted the citation before the book saw print. Copemicu in a letter to Pope Paul III: "According to Cicero, Nicetas had thought the Earth was moved ... According to Plutarch [who discussed tarchus]... certain others had held the same opinion. When from therefore, I had conceived its possibility, I myself also began to upon the mobility of the Earth."

The First Scientist

By Brian Clegg

Today, if you were to ask people to identify an inventor who seemed ahead of his time, they would be likely to point to Leonardo da Vinci. Yet some 300 years earlier Roger Bacon had already envisaged many of the devices that Leonardo was to describe and draw.

In his letter De mirabile potestate artis et natura ('On the Marvellous Power of Art and Nature'), Bacon set out a veritable catalogue of wonders. Because of his constant concern to separate the realities of nature and the fictions of magic, he began by establishing a very clear context for these inventions. These are 'marvels wrought through the agency of Art and of Nature … In these there is no magic whatsoever because, as has been said, all magical power is inferior to these works and incompetent to achieve them.'

Bacon then launched into an incredible list of modern concepts. He made the extraordinary claim that these devices had been constructed in antiquity and in his own time. He claimed to be 'acquainted with them explicitly, except with the instrument of flying which I have not seen. And I know no one who has seen it. But I know a wise man who has thought out the artifice.'

There seems a remarkable honesty in the way he admits to not having seen the flying machine. Yet it is widely thought that throughout the letter Roger was in fact describing ideas of his own devising, none of which had ever become physical reality. It is just possible, though, that some of these remarkable inventions had been made real by Peter Peregrinus.[37]

So what was in Bacon's collection of medieval mechanical wonders?

> It is possible that great ships and sea-going vessels shall be made which can be guided by one man and will move with greater swiftness than if they were full of oarsmen.[38]

Ships and boats were, of course, nothing new. Prehistoric people built simple rafts consisting of logs bound together in a flat structure, or hollowed out a tree in the form of a dugout canoe. The Romans had turned the fighting ship into a sophisticated weapon of warfare. Yet the two means of motive power available to shipbuilders had not changed over all the years of construction. To make a boat move through the water took the effort of men pulling on oars or the force of the wind on its sails. It wasn't until the end of the nineteenth century, 600 years after Bacon's death, that a steam engine could be made small and light enough to fit into a ship's hull. Even then the requirement that a 'great ship' should be guided by 'only one man' would not be realized until the advent of computerized control systems in the late twentieth century.

> It is possible that a car shall be made which will move with inestimable speed, and the motion will be without the help of any living creature. Such, it is thought, were the currus falcati which the ancients used in combat.

Like the self-propelled ship, Bacon's horseless carriage required the steam engine to make it possible, but it was truly brought into existence in 1885 when Karl Benz and Gottlieb Daimler independently came up with cars that did indeed move 'without the help of any living creature' thanks to an early petrol engine. Yet the propulsion alone wasn't Bacon's only original idea. For many, a horseless carriage would simply replace a horse with a mechanical equivalent, travelling at a familiar walking pace. But Bacon's carriage was going to 'move with inestimable speed'-exactly how we might imagine a medieval observer would regard a modern vehicle.

Interestingly, Bacon seems to have arrived at this concept as a result of a misunderstanding. He thought that self-propelled cars were possible because he assumed that something similar had previously existed, in the shape of the mythical currus falcati. These vehicles were rumoured to pull themselves, but this must have been propaganda rather than fact-currus falcati are thought to have been perfectly ordinary chariots.

In coming up with an idea as a result of a misunderstanding, Bacon made himself an early entry in a list of illustrious but confused inventors. When the French king Louis XVI saw the Montgolfier brothers' hot-air balloon in 1783, he was so impressed that he ordered his scientific officer, Jacques Alexandre Cesar Charles, to develop one for him. Charles had not seen the Montgolfiers' balloon, and the king was unable to describe how it worked. Starting from scratch, Charles reasoned incorrectly that the Montgolfiers must have used the newly discovered gas hydrogen, as this was lighter than air, and so invented the gas balloon.

> It is possible that a device for flying shall be made such that a man sitting in the middle of it and turning a crank shall cause artificial wings to beat the air after the manner of a bird's flight.

The flying machine, which Bacon owns up to not having seen for himself, is the only machine where the principle described is not a practical one. Many attempts would later be made to construct flying machines with wings that beat like those of a bird, but such endeavours were doomed to failure because the effort required to lift the bulk of a person far exceeds the amount of energy that could ever be put into the wings. Birds only manage to become airborne because they are extremely light for their bulk and have a remarkable amount of power in their musculature. Toy-sized craft that use the flapping principle (ornithopters) have been made, but the principle does not scale up to full-size aircraft.

The most basic form of workable airborne device, the kite, had been around since the fifth century BC, but it was Bacon who gave a hint of the principle on which the modern fixed-winged aeroplane functions, pointing out that air should be able to support it in the same way that water supports a boat. Leonardo da Vinci was to revisit Bacon's concept of an ornithopter, but he also devised a helicopter-like mechanism and, most significantly, a glider. The first practical glider was made by the English inventor George Gayley, who also drew up plans for a fixed-winged powered aircraft in 1799, but it was not *until* 17 December 1903 that American bicycle-makers Wilbur and Orville Wright built and flew the first piloted powered flying machine, and brought Bacon's idea to life.

> Similarly it is possible to construct a small-sized instrument for elevating and

depressing great weights, a device which is most useful in certain exigencies. For a man may ascend and descend and may deliver himself and his companions from peril of prison by means of a device of small weight and of a height of three fingers and a breadth of four.

It is possible also easily to make an instrument by which a single man may violently pull a thousand men toward himself despite the opposition, or other things which are tractable.

While Bacon is not explicit about his subject here, he seems to be describing a pulley system where the effort applied is magnified by the pulleys. Simple machines for amplifying force—pulleys, levers and screws-had been common since ancient times, but the form Bacon describes would not be available until it became possible to construct complex precision pulley blocks, perhaps 500 years later. For good measure, he throws in the concept of a lift, a device that would not become reality until the nineteenth century.

It is possible also that devices can be made whereby, without bodily danger, a man may walk on the bottom of the sea or of a river. Alexander used these to observe the secrets of the sea, as Ethicus the astronomer relates.

This was probably the least original of Bacon's machines. For around 1,600 years, practical means had sought to make it possible for a human being to survive under water. As Bacon points out, Alexander the Great was said to have tried out a primitive form of diving bell, and Aristotle mentions similar devices. However, it wasn't until the eighteenth century, when there was a better under-standing of atmospheric pressure, that practical diving bells began to be produced.

Bacon's collection of wonders was amazing. *In one short burst in a letter to an acquaintance he listed self-powered ships, the horseless carriage, the flying machine, something that sounds like a pulley system, and a diving suit or diving bell, most of which would not become practical for another 600 years. Compare this with the work of the feted *twentieth-century, science-fiction* writer Sir *Arthur* G. Clarke, author *of* 2001: A Space Odyssey, who is usually cited as the leading example of technological foresight. In an article published in Wireless World in 1945, he predicted the use of communications satellites around twenty years before they became reality.[39] There is no doubt that this excellent writer had an unusually precise ability to speculate on the future of technology, but Bacon puts even Clarke into the shade.

Of course, Bacon was not alone in describing marvellous inventions. Classical romances describe all manner of impossible constructs, but they were presented as magical devices. In the mythology of ancient Greece, for example, Hephaestus, the smith of the gods, was said to have constructed a magically powered man of brass called Talus, who guarded the island of Crete. This remarkable creation may have indirectly inspired one of the totally fictional legends that by Elizabethan times had been constructed around Roger Bacon. Bacon was said to have built a head out of brass, so cleverly constructed that it could speak. It was not based on scientific principles, however, being powered by exposure to 'the continual fume of the six hottest simples' (a 'simple' was an extract from a single plant). This was just the sort of magical fantasy that Roger was so scathing about.

By contrast, the inventions he discussed were set squarely in the field of science and engineering. Remember his remark that these are 'marvels wrought through the agency of Art and of Nature … In these there is no magic whatsoever.' There was nothing obscure about them either. Compared with the cryptic and veiled predictions of a Nostradamus, Bacon's work was clear futurology, not requiring the sort of after-the-event interpretation that is so often applied to prophesy. Bacon is not indulging in fantasy, but describing what he believed human ingenuity could construct. Roger Charles, the nineteenth-century Victorian biographer, writes of this list that 'it is a dazzling picture and one to confound modern science, which believes itself born yesterday.' We are so used to the developments of the last hundred years that perhaps it takes a Victorian view to appreciate how remarkable were the concepts set out in Bacon's letter.

Once he was in full flow, Bacon did not stop with mere mechanical devices. Optics had always been a subject that fascinated him, and here too his inventive mind could see all sorts of possibilities. He began with the idea of using mirrors and lenses so that 'one appears as many, one man an army' an effect suggested could be used in war so that 'infinite terror may be cast upon a whole city or upon an army so that it will go entirely to pieces'. Whether such an illusion could ever have fooled anyone is another matter; however, he then went on to describe optical devices that would come into use hundreds of years later, where 'lenses are contrived so that the most distant objects appear near at hand and vice versa:

> We may read the smallest letters at an incredible distance, we may see objects however small they may be, and we may cause the stars to appear wherever we wish. So, it is thought, Julius Caesar spied into Gaul from the seashore and by optical devices learned the position and arrangement of the camps and towns of Brittany.

Bacon's incorrect assumption that Julius Caesar had used telescopes did not make his concept any less remarkable. After all, Bacon was the foremost expert on optics of his day. He understood as few had done before how the primitive lenses and mirrors then in existence bent light rays. Others, da Vinci included,

would follow him in referring to the possibility of making telescopes and microscopes, but the first known examples would not be made until 300 years later. It was in the late sixteenth century that the Digges father-and-son team may have first demonstrated a telescope,[40] and around 1590 when another family partnership, Hans and Zacharias Janssen, constructed a microscope.

With the telescope and microscope under his belt, Bacon went on to imagine using the power of the basilisk, a non-existent creature that was then thought to kill as a result of emanations in its sight:

> Devices may be built to send forth poisonous and infectious emanations and influences wherever a man may wish. Aristotle taught this to Alexander, so that by casting the poison of the basilisk over the walls of a city which held out against his army he conveyed the poison into the city itself.

If this sounds much more vague and mystical than a piece of modern technology, bear in mind that Bacon assumed that the basilisk was a real creature, with a natural if quite amazing way of imposing death through sight (sight was then assumed to be a flow of light emanating from the eyes). He goes on to describe channelling this deadly light using lenses:

> But of sublimer powers is that device by which rays of light are led into any place that we wish and are brought together by refractions and reflections in such fashion that anything is burned which is placed there. And these burning glasses function in both directions, as certain authors teach in their books.

Bacon went on to describe orrery-like instruments that were working models of the heavens. Such devices would not see daylight until the seventeenth century. The orrery is sometimes confused with the much older astrolabe, but that was a two-dimensional (if sometimes beautiful) instrument of metal circles used to observe the sky, the precursor of the sextant, and dates back to the ancient Greeks. Bacon's description is so precise that it is tempting to wonder if his one-time colleague Peter Peregrinus had devised such a machine, perhaps inspired by Bacon's work on geographical and celestial locations:

> The great power of mathematics can build a spherical instrument, like the artifice of Ptolemy in Almagest, in which al! heavenly bodies are described veraciously as regards longitude and latitude, but to make them move naturally in their diurnal movement is not within the power of mathematics. A faithful and magnificent experimenter might aspire to construct an instrument of such materials and of such an arrangement that it would move naturally in the diurnal motion of the heavens … In the presence of such an instrument all other apparatus of the Astrologers, whether the product of wisdom or mere vulgar equipment, would cease to count any more. The treasure of a king would scarcely merit comparison with it.

Bacon suggests that the device could be made to move directly by the action of the heavens, just as other physical phenomena, from comets to the tides, are driven by external forces. While there is no suggestion that Bacon had in mind a working mechanism, the idea of driving an 'engine' by gravity was to return in instruments as common as the humble cuckoo clock. The cuckoo clock is kept in motion by the weights that hang below, which exert a force through the action of gravity, just as the tides are produced by the gravitational pull of the Moon.

Finally, and in great depth, Bacon explored the manufacture of gunpowder. Others had described the inflammable nature of black powder, and various highly flammable compounds had been in use in warfare since Greek fire, a gelatinous petroleum-based mixture, had been developed in the seventh century AD to attack ships at sea. But Bacon was the first to put across the explosive force of gunpowder:

> For the sound of thunder may be artificially produced in the air with greater resulting horror than if it had been produced by natural causes. A moderate amount of proper material, of the size of the thumb, will make a horrible sound and violent coruscation. Such material may be used in a variety of ways, as, for instance, in a case similar to that in which a whole army and city were destroyed by means of the strategy of Gideon, who, with broken jugs and torches, and with fire leaping forth with ineffable thunder, routed the army of the Midianites with three hundred men.

There were limits to Bacon's vision. He did not conceive of the ability of gunpowder to blast a projectile, a discovery that would change the face of weaponry for ever. Even so, his was the first real insight into the military value of black powder and the wider capabilities of science to make a difference to everyday life.

With Bacon's list of inventions in front of us, it is worth exploring a little further his assertion that he had seen most of the mechanical devices himself. We know from Bacon's own account that Peter Peregrinus was a great experimenter who was prepared to try out things that others would simply not consider.[41] It is not inconceivable that Peter was capable of putting together models that really did propel themselves as Bacon describes, perhaps using weights and pulleys to power them.

Also, Bacon himself would certainly have been able to experiment with pulleys and levers. And though he does not state explicitly in the letter that he had constructed the optical devices that are mentioned, we do know that he carried out many experiments with mirrors, prisms and lenses, as he describes in the Opus majus:

> The wonders of refracted vision arc still greater; for it is easily shown by the rules stated above [demonstrating the workings of lenses] that very large objects can be made to appear small, and the reverse, and very distant objects will seem very close at hand, and conversely. For we can so shape transparent bodies, and arrange them in such a way with respect to our sight and objects of vision, that the rays will be refracted and bent in any direction that we desire, and under any angle we wish we shall see the object near or at a distance.
>
> Thus from an incredible distance we might read the smallest letters and number grains of dust and sand owing to the magnitude of the angle under which we viewed them …[42]

With his particular expertise in this field, it seems quite possible that he may not only have described the microscope and telescope but also built primitive versions during his twenty years of research.

If so, why was it another 400 years before these devices were put to practical use? Even in Bacon's day, a telescope would have had obvious military benefits, and Bacon was clearly aware of this potential in his reference to Julius Caesar spying on Gaul. Also, an ever-present motivation in his work was the desire to give the Church more strength to stand up to its enemies. If he could have presented Christianity with an unparalleled surveillance tool, why did he not do so? If he had built a telescope, why was it not pressed into use immediately?

Although this is pure speculation on my part, I think it quite possible that Bacon had constructed a simple telescope by around the late 1240s. Given his fascination with optics and the experience of experimentation he had gained while working with Peter Peregrinus, it is hard to believe that he never used two lenses to form a simple telescope. The lenses available then were very crude, giving a misty, distorted image, yet it would take little more than two such lenses and a tube of leather to make a workable instrument. However fuzzy the view, it would show things more distant than the naked eye could see, and that would have been enough to prove the telescope's worth.

But if Bacon did make a simple device, he was never to get the opportunity to show it off outside his immediate circle of friends. His life was to undergo another drastic change, and any of his belongings, including anything he had invented, would probably have been left with his family—who were to lose everything in the turmoil of civil war. If Bacon's precious devices had been stored at the family home, they would have been destroyed. And there would be good reasons why Bacon would never re-create such instruments later on.

For the moment, though, he was to about take a bizarre leap into the unknown. He was about to surrender voluntarily everything that he held dear.

25. Quoted in Catto, *History of the University of Oxford*.
26. Paris, *Historia major.*
27. Quoted in Easton, *Roger Bacon and His Search for Universal Science.*
28. Easton, *Roger Bacon and His Search for Universal Science.*
29. *De retardandis.*
30. *De retardandis,* Dedication.
31. E.g. Easton, *Roger Bacon and His Search for Universal Science.*
32. *Opus tertium.*
33. *Opus tertium.*

34. Easton, *Roger Bacon and His Search for Universal Science*.

35. *Opus tertium*.

36. Platt, *Medieval England*.

37. Easton, *Roger Bacon and His Search for Universal Science*.

38. *De mirabile* (except where indicated, all subsequent quotes in this chapter are from *De mirabile)*.

39. Clarke, 'Extra-terrestrial relays'.

40. Ronan, 'Leonard and Thomas Digges'.

41. *Opus tertium,* referred to in Westacott, *Roger Bacon in Life and Legend* and Hackett, *Roger Bacon and the Sciences*.

42. *Opus majus,* Part 5, Last distinction, Chapter IV

43. *Compendium studii philosophiae*.

44. Most of the information on the early years of the Order is drawn from Moorman, *History of the Franciscan Order*.

45. Catto, *History of the University of Oxford*.

46. Most of the information on Grosseteste is from Southern, *Robert Grosseteste*.

47. Grosseteste, *De luce*.

48. *Opus majus,* Part 5, First distinction, Chapter I.

49. *Opus majus,* Part 5, First distinction, Chapter I.

50. *Opus majus,* Part 5, First distinction, Chapter I.

51. Itard, *Les Livres arithmetique d'Euclide*.

52. For further information on Alhazen see Clegg, *Light Tears*.

53. *Opus majus,* Part 5, Seventh distinction, Chapter IV

Copernicus' Secret

By Jack Repcheck

Toward the end of the summer, Copernicus and Rheticus left Giese in Lubawa and returned to Frombork. There Rheticus put the final touches on his manuscript about Copernicus's theory of the heavens, the closing words being, "From my library at Frombork, September 23, 1539." It is not known where Rheticus's library was, but it was most likely a room in the canon's house. After finishing the manuscript, he and Heinrich Zell traveled to Gdansk, the largest city in the area and probably the only one with a printer. Rheticus gave the stack of paper to Franciscus Rhodes to print and publish. It appears that Zell stayed in Gdansk to supervise the process and proofread the pages.

Rhodes must have been an efficient printer because the book was ready for sale by April of 1540. Thus, it took only eleven months from the moment when Rheticus knocked on Copernicus's door on that memorable day the previous May until there was at last a description of the science of the Canon of Frombork. Those intellectuals who had been waiting for it were not disappointed—the *Narratio prima* was a splendid work. Though written quickly and then published without delay, it was nonetheless smoothly composed and riveting to read. To this day, it is arguably the best primer on the heliocentric theory of Copernicus. The *Narratio* was doubly impressive because it was not a straight summary of the canon's larger work. Instead, Rheticus was able to absorb the wealth of detailed and technical information in Copernicus's manuscript and then construct his own narrative in order to address the skepticism that he knew would greet the theory.

The short book (in English it is about 100 pages long, significantly longer than the *Commentariolus*) was written in Latin as a long and formal letter addressed to "The Illustrious Johann Schoner, as to his [Rheticus's] own revered father/' Though Schoner may not have urged the young Lutheran to travel all the way to Frombork to seek Copernicus, it was Schoner's fascination with the canon that had planted the seed in Rheticus's mind. Rheticus continues: "I promised to declare as early as I could, whether the actuality answered to report and to my own expectations ... To fulfill my promises at last and gratify your desires, I shall set forth ... the opinions of my teacher/' Rheticus does not refer to Copernicus by name; he is called "my teacher" throughout (though the canon's name was given on the book's title page).

The Wittenberg professor early on gives Copernicus the highest compliment that could be bestowed on an astronomer, and thus prepares the reader for the significance of what is to follow:

First of all, I wish you to be convinced, most learned Schoner, that this man whose work I am now treating is in every field of knowledge and in mastery of astronomy not inferior to Regiomontanus. I rather compare him with Ptolemy . ... My teacher has written a work of six books in which, in imitation of Ptolemy, he has embraced the whole of astronomy, stating and proving individual propositions mathematically and by the geometric method.

Rheticus is careful throughout his book to compliment Ptolemy, Aristotle, and all the ancients. He never describes Copernicus as a revolutionary—rather Rheticus shows him as a respectful astronomer who assiduously studied every available source and was committed to the evidence. He also makes many references to God as the divine creator of the heavens, to the Roman gods, to Aesop's fables, and to Regiomontanus. Although there are several passages about astrology, the zodiac, and horoscopes, the focus of the book is on Copernicus's theory of the movements of the heavenly bodies.

Although the *Narratio* is a summary of Copernicus's manuscript, Rheticus does not discuss heliocentrism and the moving earth right away. He begins with a brief on the motion of the fixed stars, stressing the importance of the observations Copernicus made in Italy and also one he made in 1525. Next he discusses the tropical year, which was critical information for calendar reform. Toward the end of this section, Rheticus gives the first hint of what is to come: "That there necessarily was a deficiency of 19/20 of a day from Hipparchus to Ptolemy, and from Ptolemy to Albategius of about 7 days, I have deduced, not without the greatest pleasure, most learned Schoner, from the foregoing theory of the motions of the stars and from my teacher's treatment of the motion of the sun, as you will see a little further on."

Then, a dozen pages into the *Narratio*, Rheticus displays his astrological side.

I shall add a prediction. We see that all kingdoms have had their beginnings when the center of the eccentric was at some special point on the small circle. Thus, when the eccentricity of the sun was at its maximum, the Roman government became a monarchy; as the eccentricity decreased, Rome too declined, as though aging, and then fell. When the eccentricity reached the boundary and quadrant of mean value, the Mohammedan faith was established; another great empire came into being and increased very rapidly, like the change in the eccentricity. A hundred years hence, when the eccentricity will be at its minimum, this empire too will complete its period. In our time it is at its pinnacle from which equally swiftly, God willing, it will fall with a mighty crash. We look forward to the coming of our lord Jesus Christ when the center of the eccentric reaches the other boundary of mean value, for it was in that position at the creation of the world. This calculation does not differ much from the saying of Elijah, who prophesied under divine inspiration that the world would endure only 6,000 years, during which time nearly two revolutions are completed. Thus it appears that this small circle is in very truth the Wheel of Fortune, by whose turning the kingdoms of the world have their beginnings and vicissitudes. For in this manner are the most significant changes in the entire history of the world revealed, as though inscribed upon this circle.

Curiously, just one year after the *Narratio* appeared, Martin Luther prepared a chronology of the world in which he predicted the end of the earth after 6,000 years and cited the same prophesy by Elijah.

About one-quarter of the way into the short book, Rheticus finally clears his throat and begins to discuss the highlight of Copernicus's work. He reluctantly states that Ptolemy's model cannot explain the movements of celestial bodies in a consistent way, so " [i]t was therefore necessary for my teacher to devise new hypotheses …" A few pages later, the heliocentric theory is revealed for the first time in a published book:

> The planets are each year observed as direct, stationary, retrograde, near to and remote from the earth, etc. These phenomena, besides being ascribed to the planets, can be explained, as my teacher shows, by a regular motion of the spherical earth; that is by having the sun occupy the center of the universe, while the earth revolves instead of the sun on the eccentric, which it has pleased him to name the great circle. Indeed, there is something divine in the circumstance that a sure understanding of celestial phenomena must depend on the regular and uniform motion of the terrestrial globe alone.

Rheticus goes on to discuss the second stunner in Copernicus's model—"the earth, like a ball on a lathe, rotates from west to east, as God's will ordains; and that by this motion, the terrestrial globe produces day and night…"

In mid-February 1540, a batch of preliminary pages from the Narratio was sent to Philipp Melanchthon in Wittenberg. No doubt Rheticus wanted to inform the University of Wittenberg's rector about his activities, and he also knew that the several references to astrology would likely please Melanchthon. Another set of advance pages was sent to Andreas Osiander in Nuremberg.

Osiander wrote back to Rheticus immediately. He was clearly captivated by what he had read. After a lengthy comment on Rheticus's passage in his book about the second coming of Christ and the determining of the precise age of the earth, he says, "Yet enough of this, of that above, just as I ask you again and again that you offer me your friendship, I ask that you apply yourself diligently and win over the friendship of that man [Copernicus] for me as well. I don't risk writing him at the present, and although I didn't intend to, you will certainly keep these my triflings from him."

Rheticus and Giese were determined to use the Narratio as advance publicity for their mentor and friend. Immediately after obtaining their first copies of the complete book, Rheticus sent several to his colleagues in Nuremberg—certainly Schoner, Osiander, and Petreius—and one to Gasser in Feldkirch. Giese sent a copy to Albert, the duke of Prussia.

The reaction was immediate and emphatic. Gasser wrote to a friend sometime in mid-1540:

> The book may differ from the manner of teaching practiced so far. As a whole it may appear to run contrary to the usual theories of the schools and even to be (as the monks would say) heretical. Nevertheless, what it undoubtedly seems to offer is the restoration—or, rather, the rebirth—of a true system of astronomy. For in particular it makes highly evidential claims concerning questions that have long occasioned much perspiration and debate across the world not only by very learned mathematicians but also by the greatest philosophers: the number of the heavenly spheres, the distance of the stars, the rule of the Sun, the position and courses of the planets, the exact measurement of the year, the specification

of solstitial and equinoctial points, and finally the position and motion of the earth itself.

In August of that year, Johannes Petreius, the Nuremberg publisher, did a most unusual thing. He dedicated a book to Rheticus in the form of a letter, and in that letter he boldly asks Rheticus to convince Copernicus to publish his long-awaited book and to publish it with him, Petreius, in Nuremberg. After congratulating Rheticus on the Narratio (calling it a "splendid description"), he states that "I consider it a glorious treasure if some day through your urging his observations will be imparted to us." Petreius hoped that Rheticus would see the dedication as a "kind of reward from us for your labors and study." Then to close the deal, Petreius finishes his letter by reminding Rheticus how Nuremberg is a major trading hub, that his publishing company is able to distribute books to every corner of Europe, and that Schoner, the scholar who had taken such good care of Rheticus during his stay and had taught him much, also wants to see Copernicus published in Nuremberg: "It will fall on you, not only to commend our service, but also to acknowledge and proclaim the great favor of Schoner toward you."

The book that Petreius dedicated to Rheticus was by Antonius de Montulmo, a fourteenth-century physician, entitled On Natal Horoscopes. It was on Regiomontanus's Index of Books, and Schoner had discovered the additions to the work that Regiomontanus had planned to include, so it was a very significant book. Dedicating it to the young Wittenberg professor, who was known to be fascinated with horoscopes, represented a serious lobbying effort. Petreius meant business—he was determined to be Copernicus's publisher.

The letters from Osiander, Gasser, and especially Petreius were persuasive. Gasser was a knowledgeable astrologer and published author. He represented the learned community of philosophers whom Copernicus hoped to impress with his work. Petreius's letter was at least as important to the canon because it meant that the astronomers of Nuremberg—which included Schoner and Osiander—all of whom published with Petreius, were also eager to read Copernicus's book. With this level of support, Copernicus finally relented. He would at last publish his entire manuscript, thus allowing anyone who could afford to buy the book or borrow it the chance to study the details of his revolutionary theory of the celestial bodies. On July 1, 1540, he wrote to Andreas Osiander, asking for his advice about how to stifle or at least minimize the opposition to his radical system that was likely to greet his book.

So, starting in the midsummer of 1540, Copernicus and Rheticus went to work preparing the "larger work" for publication.

COPERNICUS's OLD and dog-eared manuscript was hundreds of handwritten pages long. Over the next twelve months, Copernicus went back through the massive work and made corrections, brought it up to date, and otherwise revised it for publication. He did most of this work himself. One of the three books that Rheticus had given Copernicus the previous year, Regiomontanus's On Triangles of Every Kind, was particularly significant, because it caused Copernicus to go back through his trigonometry section and make many substantive revisions.

Because the manuscript that Petreius would later work from was written in a hand different from Copernicus's, scholars have long assumed that Rheticus took the finished chapters from Copernicus and redrafted them to create a clean copy for the publisher. (Incredibly, Copernicus's original manuscript has survived the nearly five centuries since its completion. Copernicus never parted from the original; it is now in the archives at the University of Krakow.) While waiting for Copernicus to finish his individual chapters, Rheticus and Zell found a way to occupy themselves—they finished the detailed map of Prussia that had been started many years earlier by Copernicus and Alexander Scultetus. Thus, the two Lutherans were seen traveling all over the region surveying the countryside.

In late 1540, probably November or December, Rheticus returned to Wittenberg to teach the Introductory Astronomy-Astrology course. It was a very short visit, after which it appears that he returned immediately to Frombork.

A couple of distractions intruded on Copernicus's efforts to finish his manuscript. In January 1541, the trial commenced over who had the right to take over Alexander Scultetus's house, vacant since the suspected heretic had fled Frombork several months earlier. Recall that at the trial, which was presided over by a panel of canons, one of the litigants demanded that Copernicus and Niederhoff be forced off the panel because they had been accused of the same offense as Scultetus—that is, Lutheran heresy. The hearing to address the complaint about Copernicus and Niederhoff was put off until April.

Fortunately, Copernicus was able to avoid the embarrassment of the inquiry when in early April he was summoned to Krolewiec to attend to the ill George von Kunheim, one of the duke of Prussia's advisors. A new judicial investigation was rescheduled for June. That month came and went without any record of a follow-up hearing.

In July 1541, Achilles Gasser sent Rheticus a copy of the second edition of the Narratio prima, which he had personally supervised in Basel. Gasser related in the letter enclosed with the volume that "a stream of requests" was now coming for Copernicus to publish the larger work.

Dantiscus himself received two letters in July from correspondents in Brussels and Louvain urging him to do everything in his power to see to it that Copernicus's book was published expeditiously The first was from one Cornelius Scepperus, and he stated that the Narratio "has made the name of Copernicus famous" The second was from the Dutch scholar Gemma Frisius. In his letter he related to Dantiscus that because of Copernicus and Rheticus, Warmia had become the new intellectual hot spot, "the shelter of the Muses' and their work on the heavens had gained them new admirers.

Nicolaus Copernicus revised the last page of his decades-in-the-making manuscript in late summer of 1541. He had now done everything he could. The rest of the process would be the responsibility of others. Rheticus packed the "fair" copy, the one that he had copied from Copernicus, in his belongings and left Frombork in September, ending a twenty-five-month stay. Many years later he told one of his patrons, "After I had spent about three years in Prussia, the great old man charged me to carry on and finish what he, prevented by age and impending death, was himself unable to complete."

The Publication

Rheticus could not Miss another semester at the University of Wittenberg. So in late September 1541, he and Zell said good-bye to Copernicus and began their journey back to Wittenberg. Rheticus had spent over two years in Warmia, and before that eight months in the Nuremberg area. Except for his brief return to Wittenberg in the winter of 1540–41, he had been away from his home university for almost three years. Rheticus should have returned as a conquering hero—the author of a much talked-about book, and the lone disciple of a visionary astronomer. And, he possessed the most acclaimed unpublished manuscript in Europe.

His reception in Wittenberg was mixed, though. Back in June 1539, just after Rheticus had arrived in Frombork, Luther said at one of his dinner seminars:

> There is mention of a certain new astrologer who wanted to prove that the earth moves and not the sky, the sun, and the moon. This would be as if somebody were riding on a cart or in a ship and imagined that he was standing still while the earth

and the trees were moving … So it goes now. Whoever wants to be clever must agree with nothing that others esteem … I believe the holy scriptures, for Joshua commanded the sun to stand still and not the earth.

Melanchthon's attitude was probably a little more problematic for Rheticus because he was Rheticus's boss and had been the one to send him away from Wittenberg to become a better astrologer in the first place. Melanchthon wrote to a correspondent in October 1541, just as Rheticus had resumed teaching: "Many hold it for an excellent idea to praise such an absurd matter, like that sarmatic [Polish] astronomer, who moves the earth and lets the sun stand still."

Most of his colleagues, however, must have been happy to have Rheticus back. They recognized the dramatic contribution to astronomy that he had made with the Narratio prima. No sooner had Rheticus unpacked than he was told that he had been elected by his fellow professors as their new dean of faculty. The election as dean indicated that Melanchthon still supported Rheticus, even though he disagreed with the science Rheticus had learned at the feet of the "sarmatic astronomer."

Since he was teaching again in Wittenberg, Rheticus was unable to make the long journey to Nuremberg to personally deliver the manuscript for On the Revolutions to the publisher Petreius. While in Wittenberg, though, he took the unusual step of taking the two trigonometry chapters from Copernicus's manuscript and publishing them as a separate short book. This book, entitled On the Sides and Angles of Triangles, was published in the late winter or early spring of 1542, and it was essentially a complement to Regiomontanus's On Triangles of Every Kind. The title page identified the author, in Latin, as "D. [Doctor] Nicolao Copernico Toronensi [from Toruri]." But Rheticus did not just lift the chapters from Copernicus's manuscript. He reworked some of

the material and improved it. He dedicated the short book to Georg Hartmann, a Nuremberg mathematician whom he had befriended along with Schoner in 1538–39. In the preface Rheticus announced, "There has been no greater human happiness than my relationship with so excellent a man and scholar as [Copernicus] is. And should my own work ever make any contribution to the general good (to the service of which all our efforts are directed), it shall be owing to him."

When the spring semester ended in May 1542, Rheticus finally left for Nuremberg. He had been holding Copernicus's manuscript for eight months, but once he was in the city of Schoner, the pace picked up immediately. Petreius was given the manuscript and got to work right away Surprisingly, Rheticus left Nuremberg for Feldkirch to visit his mother and Gasser barely a month later. Perhaps he realized that Petreius did indeed know how to set a technical book.

Meanwhile, the drumbeat of anticipation had already started. Rheticus's colleague at Wittenberg, Erasmus Reinhold, who held the chair in higher mathematics, mentioned On the Revolutions in the preface to his new commentary on Peur-bach's New Theory of the Planets, a much-anticipated book that appeared in 1542. He announced that it was expected that this astronomer from Prussia was the next Ptolemy.

On June 29, a letter was sent by a resident of Nuremberg to a friend, in which he delivered the following news:

Prussia has given us a new and marvelous astronomer, whose system is already being printed here, a work of approximately a hundred sheets length, in which he asserts and proves that the Earth is moving and the stars are at rest. A month ago I saw two sheets in print; the printing is being supervised by a certain Magister [professor] from Wittenberg.

The same sheets that the writer of the letter saw were also sent to Copernicus, so he knew that work on the book had finally begun. At this point he wrote the last piece to be included in the book, which was his preface and introduction.

The summer of 1542 was the last time that everything went well. First, Rheticus accepted a new position at the University of Leipzig. Leipzig was a fine university and his new job represented a promotion to professor of higher mathematics, the job that Reinhold held at the University of Wittenberg. Rheticus also received a substantial raise. In addition, his new boss was Joachim Camerarius, Melanchthon's best friend. So, the move was surely endorsed by the rector. Melanchthon did however caution Camerarius to spell out clearly Rheticus's obligations and salary, leaving nothing to interpretation. Melanchthon had not forgotten the one-year leave that had turned into three years. The new professor of higher mathematics had to depart Nuremberg for Leipzig in mid-October, 1542.

Before Rheticus left, he and Petreius decided that the book still needed an expert overseer. The person asked to assume this role was Andreas Osiander, the theologian and philosopher who had come to know Rheticus in 1538–39, and who had taken such an intense interest in Copernicus and his theory once he read the Narratio. He had written or edited five books with Petreius since 1540, so he was a logical choice.

Recall that Osiander had corresponded with Copernicus in 1541, and that Copernicus had specifically asked for his suggestion as to how to minimize the uproar that might await his book. Osiander had answered that one way might be to present the underlying theory in the book as mere hypothesis and essentially to say to the reader, "Do not worry so much about the theory, it's really just the results that matter/' Copernicus had rejected this idea. But Osiander still held strong to his opinion. It seems to have been in his character—he was so stubborn in his theological beliefs that he had burned many bridges by this time

and was marginalized by the leading Lutherans in Nuremberg. He was certain that a statement like the one he endorsed should begin Copernicus's book.

So Osiander took advantage of his position as the overseer of the publication process and clandestinely slipped in a one-page "To the reader . .." preface to the book. Unfortunately, it was the first thing that the reader encountered. Also, it was anonymous. Read carefully one could tell that it had not been written by Copernicus, but most readers would have assumed that it was written by the author. Worse, Osiander went too far. It was not incorrect to call Copernicus's theory a hypothesis, but Osiander made other assertions:

> For this art [astronomy] is completely and absolutely ignorant of the causes of the apparent nonuniform motions. And if any causes are devised by the imagination, as indeed very many are, they are not put forward to convince anyone that they are true, but merely to provide a reliable basis for computation … Therefore alongside the ancient hypotheses, which are no more probable, let us permit these new hypotheses also to become known … So far as hypotheses are concerned, let no one expect anything certain from astronomy, which cannot furnish it, lest he accept as the truth ideas conceived for another purpose, and depart from this study a greater fool than when he entered it.

At about the same time that Osiander was misrepresenting *On the Revolutions,* another disaster struck. Sometime before December 8,1542, Copernicus suffered a debilitating stroke. On that day, Giese responded to a letter sent by a Frombork canon, "I was shocked by what you wrote about the impaired health of the venerable old man, our Copernicus." His condition was confirmed a few weeks later when Dantiscus responded to Gemma Frisius in the Netherlands, who was very eager to read Copernicus's book, that the canon was now paralyzed. He went on to mention the book itself was being looked after at the publisher's by Rheticus.

The long-labored-on and long-awaited magnum opus of Nicolaus Copernicus finally rolled off Petreius's presses sometime before the end of March 1543. Yet, instead of wild jubilation, the four people closest to the project had horrible surprises awaiting them.

First, Giese. He was so appalled by the anonymous preface by Osiander that he could not enjoy the moment. He later wrote a long letter to Rheticus that reads in part:

> On my return from the royal wedding in Krakow, I found the two copies, which you had sent, of the recently printed treatise of our Copernicus … However, at the very threshold I perceived the bad faith and, as you correctly label it, the wickedness of Petreius, which produced in me an indignation more intense than my previous sorrow. For who will not be anguished by so disgraceful an act, committed under the cover of good faith.

Second, Petreius. One of the leading publishers in Europe, he had worked diligently to publish a very complicated book quickly and well. But shortly after the publishing process got started the main supervisor, Rheticus, left for his new position in Leipzig. Then the author himself became physically and mentally incapacitated., Yet, Petreius still got the book out, but among the first responses was the bitter reaction of Giese over the anonymous preface. Giese, a powerful man, wrote a formal letter of complaint to the Nuremberg city council, demanding that they reprimand Petreius and then force him to republish the book. Giese's letter led to a formal hearing in which

Petreius was found innocent of any wrongdoing, but which clearly caused him anguish and embarrassment. Petreius wrote an angry reply to Giese that had to be edited by the council to excise the many intemperate passages.

Third, Rheticus. Poor Rheticus suffered two depressing shocks. When he departed from Nuremberg in September or October and left the supervisory process in the good hands of Osiander and Petreius, he assumed that he had nothing more to worry about.

But this account, written many years later, gives some insight into Rheticus's reaction when he excitedly opened his copy of the book:

> Concerning this letter ["To the reader…"], I found the following words written somewhere among Philip Apian's books (which I bought from his widow) … "On account of this letter, Georg Joachim Rheticus, the Leipzig professor and disciple of Copernicus, became embroiled in a very bitter wrangle with the printer. The latter asserted that it had been submitted to him with the rest of the treatise. Rheticus, however, suspected that Osiander had prefaced it to the work, and declared that, if he knew this for certain, he would sort the fellow out in such a way that he would mind his own business and never again dare to slander astronomers. Nevertheless, Apian told me that Osiander had openly admitted to him that he had added this as his own idea."

As bad as the first shock was, the second one must have been even more discouraging. After risking arrest and the loss of his job just to visit Copernicus in the first place, then working with the astronomer for more than two years, then writing the potentially controversial *Narratio prima*, then setting up the publication of the big book through his connections in

Nuremberg—after all that, when Rheticus opened the finished book, got past Osiander's blasphemous paragraphs, and finally read Copernicus's opening words, his acknowledgments, Rheticus must have been stunned to read that although Copernicus thanked several people, he somehow forgot to thank him. This had to have been a devastating blow to the young mathematician.

Historians of science have been at pains to explain what happened. Several have said that Copernicus was probably just trying to protect the Lutheran professor, who was already in hot water with Luther and Melanchthon. But this explanation is unlikely. First, Rheticus had attached his name to the well-read Narratio prima. And he had published Copernicus's short book on trigonometry, in which Rheticus was identified, too. Finally, Tiedemann Giese, who knew Copernicus better than anyone else, had no explanation for the oversight. In the letter written in the summer of 1543, after he had finally book, Giese wrote that "your teacher failed to mention his Preface to the treatise.. I explain this oversight not by his disrespect for you, but by a certain apathy and indifference (he was inattentive to everything which was nonscientific) especially when he began to grow weak. I am not unaware how much he used to value your activity and eagerness in helping him." But Copernicus had been strong enough to remember others in the preface, so this explanation seems unpersuasive.

What happened? Was Copernicus possibly upset by the separate publication of his trigonometry chapters, containing changes by Rheticus, without his permission? Or, perhaps the delay of nearly a year before the printing began? It must have been something specific, because the oversight is glaring. Another possible explanation—and this is only speculation—is that after Copernicus observed the acclaim bestowed on the Narratio prima, and after the young and enthusiastic Rheticus left Frombork with his masterpiece, Copernicus might have sensed that he would not be around to enjoy the moment of victory,

and Rheticus surely would. Perhaps this bothered him so much that he deliberately slighted Rheticus.

When On the Revolutions rolled off Petreius's press, the canon was lying in his house in Frombork, paralyzed and dying. He may have been sufficiently alert to realize that he would not live to enjoy the acclaim that was likely to greet the first book in nearly 1,400 years to rival Ptolemy's Almagest. There must have been many times while he was actively working on the manuscript that he dreamed of the recognition that the book would bring. And he certainly must have felt excitement while preparing it for publication in 1540 and 1541. Now, knowing that the end was near, what despair he must have felt.

AFTER THE ANONYMOUS and confusing first page written by Osiander, the next item in the book is the letter of 1536 from Cardinal Schonberg, in which he urged the canon to let him publish the work at his (or the Church's) own expense in Rome. Obviously, Copernicus wanted to make it clear that Church leaders were not opposed to what was about to follow. Next, the first words of Copernicus appear, in the "Preface and Dedication to Pope Paul III.' In the preface Copernicus boldly asserts his theory that the earth moves around the sun, but he also states that the fear of being "hooted off the stage drove me to almost abandon a work already undertaken." He then thanks Cardinal Schonberg and Tiedemann Giese, but not Rheticus.

The astronomer proceeds to state that the main problem with past theories and the reason why he tried to come up with a new model was that the other ones "contradict the first principles of regularity of movement." Then he provides a great metaphor, perhaps drawing on his training as a doctor: he compares past theories as consisting of all the parts of a body, "hands, feet, head, and the other limbs"—but put together in such a way that the result is a "monster rather than a man." The author's new conception makes the parts fit as a whole. He goes on to point out that only mathematicians will really be able to pass judgment on what follows.

On the Revolutions consists of six "books." The books are composed of many short chapters. The work is carefully organized, and Copernicus took pains to provide good transitions, introductions, conclusions, and passages meant to help the reader know what has already been covered and what is coming next. But the book is unapologetically technical, with page after page of math, numerous complicated drawings, and many dense tables of numbers.

Book One is a general introduction to the model. Copernicus first discusses the importance of astronomy and then he begins his presentation. The universe is a sphere, as is the earth, the movements of the celestial bodies are regular and circular, and the earth, too, has a circular motion. All of the heavenly bodies move with uniform speed, which is a critical component of his model. Very early in the book, no more than twenty pages in, he describes the "movement of the earth"—that is, that our planet rotates once every twenty-four hours, that it revolves around the sun once every 365 days.

Book Two discusses the rotation of the earth itself and the angle of inclination of the axis. Within this section, the author describes how to construct an astrolabe, which is used to examine the position of the moon and stars. Copernicus points out that the earth's rotation and revolution are slow and natural, and that is why the planet does not break apart and the atmosphere does not blow away. Book Three addresses the movement of the earth around the sun. The remaining three books describe the movements of the moon and the other planets.

There was no doubt that Nicolaus Copernicus's book was a remarkable achievement. That much was obvious to almost every reader. But, because of its complexity, not much else was. Most interested readers would need some help to understand its implications.

Aladdin's Lamp

By John Freely

Kepler was born on 27 December 1571 in Weil der Stadt in southwestern Germany. His father was an itinerant mercenary soldier, his mother a fortune-teller who at one point was accused of being a witch and almost burned at the stake. The family moved to the nearby town of Lemberg, where Kepler was enrolled in one of the excellent Latin schools founded by the Duke of Wurttemberg. His youthful interest in astronomy had been stimulated by seeing the comet of 1577 and a lunar eclipse in 1580.

In 1589 Kepler entered the University of Tubingen, where, in addition to his studies in mathematics, physics, and astronomy, he was influenced by Platonism, Pythagoreanism, and the cosmological ideas of Nicholas of Cusa. His mathematics lectures were based on the works of Euclid, Archimedes, and of Perge. (As Kepler later said, "How many mathematicians are there, who would toil through the *Conics* of Apollonius of Perge?")

Kepler was particularly influenced by his professor of astronomy, Michael Maestlin, from whom he first learned of the heliocentric theory. In the introduction to his first book, the *Mysterium Cosmographicum,* Kepler wrote of his excitement on discovering the work of Copernicus, which he described as "a still unexhausted treasure of truly divine insight into the magnificent order of the whole world and of all bodies."

Kepler received his master's degree at Tubingen in 1591, after which he studied theology there until 1594, when he was appointed a teacher of mathematics at the Protestant seminary in the Austrian town of Graz. A year after his arrival in Graz, Kepler came up with an idea that he thought explained the arrangement and order of the heliocentric planetary system. He had learned from his reading of Euclid that there are five and only five regular polyhedra, the so-called Platonic solids, in which all of the faces are equal as well as equilateral—the cube, tetrahedron, dodecahedron, icosahedron, and octahedron—and it occurred to him that they were related to the orbits of the earth and the five other planets. He explained the scheme in his treatise the *Mysterium Cosmographicum,* published in 1596, in which his values for the relative radii of the planetary orbits agree reasonably well with those determined by Copernicus, though there was no physical basis for his theory.

> The earth's orbit is the measure of all things; circumscribe around it a dodecahedron, and the circle containing it will be Mars; circumscribe around Mars a tetrahedron, and the circle containing this will be Jupiter; circumscribe around Jupiter a cube, and the circle containing this will

be Saturn. Now inscribe within the earth an icosahedron, and the circle contained in it will be Venus; inscribe within Venus an octahedron, and the circle contained in it will be Mercury. You now have the reason for the number of planets.

Kepler sent copies of his treatise to a number of scientists, including Galileo Galilei (1564–1642). In his letter of acknowledgment, dated 4 August 1597, Galileo congratulated Kepler for having had the courage, which he himself lacked, to publish a work supporting the Copernican theory.

Kepler wrote back to Galileo on 13 October 1597, encouraging him to continue supporting the Copernican theory. "Have faith, Galilii, and come forward!" he wrote. "If my guess is right, there are but few of the prominent mathematicians of Europe who would wish to secede from us: such is the power of truth."

Galileo was born in Pisa on 15 February 1564 to a Florentine family; they moved back to Florence in 1574. He enrolled in the school of medicine at the University of Pisa in 1581, studying physics and astronomy under Francesco Buonamici, who based his teachings on Aristotle. Galileo left Pisa without a degree in 1585 and returned to Florence, where he began an independent study of Euclid and Archimedes under Ostilio Ricci.

In 1583 Galileo made his first scientific discovery, that the period of a pendulum is independent of the angle through which it swings, at least for small angles. Three years later he invented a hydraulic balance, which he described in his first scientific publication, *La Balancetta* (The Little Balance), based on Archimedes' principle, which he also used in determining the centers of gravity of solid bodies.

Galileo was appointed professor of mathematics in 1589 at the University of Pisa, where he remained for only three years. During this period he wrote an untitled treatise on motion now referred to as *De Motu* (On Motion), which remained unpublished during his lifetime. The treatise was an attack on Aristotelian physics, such as the notion that heavy bodies fall more rapidly than light ones, which Galileo is supposed to have refuted by dropping weights from the leaning tower of Pisa. Through his study of balls rolling down an inclined plane, he found that the distance traveled was proportional to the square of the elapsed time, one of the basic laws of kinematics. He also concluded that a ball rolling on a frictionless horizontal surface would continue to roll with constant velocity, while one at rest would remain motionless, thus stating the law of inertia.

In 1592 Galileo was appointed to the chair of mathematics at the University of Padua, where he remained for eighteen years. During that period he wrote several treatises for the use of his students, including one that was first published in a French translation in 1634 under the title *Le Meccaniche*, a study of motion and equilibrium on inclined planes that further developed the ideas he had presented in *De Motu*.

In May 1597 Galileo wrote to a former colleague at Pisa defending the Copernican theory. Three months later he received a copy of *Mysterium Cosmographicum*, which led to his first correspondence with Kepler.

Kepler had also sent a copy of the *Mysterium Cosmographicum* to Tycho Brahe, who received it after he had left Denmark for Germany. Tycho responded warmly, calling the treatise "a brilliant speculation," beginning a correspondence that eventually led Kepler to accept Tycho's invitation to join him at his new observatory outside of Prague. As Tycho wrote in response to Kepler's letter of acceptance: "You will come not so much as a guest but as a very welcome friend and highly desirable participant and companion in our observations of the heavens."

Kepler finally arrived in Prague with his family early in 1600, beginning a brief but extraordinarily fruitful collaboration with Tycho. When Kepler began work at Prague he had hopes that he could take Tycho's data and use it directly to check his own planetary theory. But he was disappointed to find that most of Tycho's data was still in the form of raw observations, which first had to be subjected to mathematical analysis. Moreover, Tycho was extremely possessive of his data and would not reveal any more of it than Kepler needed for his work.

These and other disagreements with Tycho led Kepler to leave Prague in April of that year, though he returned in October after considerable negotiation concerning the terms of his employment. Tycho then assigned Kepler the task of analyzing the orbit of Mars, which until that time had been the responsibility of his assistant Longomontanus, who had just resigned. Kepler later wrote, "I consider it a divine decree that I came at exactly the time when Longomontanus was busy with Mars. Because assuredly either through it we arrive at the knowledge of the secrets of astronomy or else they remain forever concealed from us."

Mars and Mercury are the only visible planets with eccentricities large enough to make their orbits significantly different from perfect circles. But Mercury is so close to the sun that it is difficult to observe, leaving Mars as the ideal planet for checking a mathematical theory, which is why Kepler was so enthusiastic about being able to analyze its orbit.

Early in the autumn of 1601 Tycho brought Kepler to the imperial court and introduced him to the emperor Rudolph II. Tycho then proposed to the emperor that he and Kepler compile a new set of astronomical tables. With the emperor's permission, this would be named the *Rudolfine Tables,* and since it was to be based on Tycho's observations it would be more accurate than any done in the past. The emperor graciously consented and agreed to pay Kepler's salary in this endeavor.

Soon afterward Tycho fell ill, and after suffering in agony for eleven days, on 24 October 1601, he died. On his deathbed he made Kepler promise that the *Rudolfine Tables* would be completed, and he expressed his hopes that it would be based on the Tychonic planetary model. As Kepler later wrote of Tycho's final conversation with him: "Although he knew I was of the Copernican persuasion, he asked me to present all my demonstrations in conformity with his hypothesis."

Two days after Tycho's death Emperor Rudolph appointed Kepler as court mathematician and head of the observatory in Prague. Kepler thereupon resumed his work on Mars, now with unrestricted access to all of Tycho's data. At first he tried the traditional Ptolemaic methods—epicycle, eccentric, and equant—but no matter how he varied the parameters the calculated positions of the planet disagreed with Tycho's observations by up to 8 minutes of arc. His faith in the accuracy of Tycho's data led him to conclude that the Ptolemaic theory of epicycles, which had been used by Copernicus, would have to be replaced by a completely new theory, as he wrote: "Divine Providence granted us such a diligent observer in Tycho Brahe, that his observations convicted this Ptolemaic calculation of an error of eight minutes; it is only right that we should accept God's gift with a grateful mind. ... Because those eight minutes could not be ignored, they alone have led to a total reformation of astronomy."

After eight years of intense effort, Kepler was finally led to what are now known as his first two laws of planetary motion. The first law is that the planets travel in elliptical orbits, with the sun at one of the two focal points of the ellipse. The second law states that a radius vector drawn from the sun to a planet sweeps out equal areas in equal times, so that when the planet is close to the sun it moves rapidly and when far away it goes slowly. These two laws, which appeared in Kepler's *Astronomia Nova* (The New Astronomy), published in 1609, became the basis for his subsequent work on the *Rudolfine Tables.*

Kepler's first two laws of planetary motion eliminated the need for the epicycles, eccentrics, and equants that had been used by astronomers from Ptolemy to Copernicus. The passing of this ancient cosmological doctrine was noted by Milton in Book VIII of *Paradise Lost,* where he describes the debate between the two world systems, Ptolemaic and Copernican.

> *Hereafter, when they come to model Heaven,*
> *And calculate the stars; how they will wield*
> *The mighty frame; how build, unbuild, contrive*
> *To save appearances; how gird the sphere*
> *With centric and eccentric scribbled o'er,*
> *Cycle and epicycle, orb in orb.*

Kepler wrote two other works on his researches before the publication of his *Astronomia Nova.* The first was the *Appendix to Witelo,* published in 1604, which dealt with optical phenomena in astronomy, particularly parallax and refraction, as well as the annual variation in the size of the sun. The second book was occasioned by another new star, which appeared in October 1604 in the vicinity of Jupiter, Saturn, and Mars. Kepler published an eight-page tract on the new star in 1606 entitled De Stella *Nova,* with a subtitle describing it as "a book full of astronomical, physical, metaphysical, meteorological, astrological discussions, glorious and unusual." At the end of the tract Kepler speculated on the astrological significance of the new star, saying that it might be a portent of the conversion of the American Indians, the downfall of Islam, or even the second coming of Christ.

Meanwhile, the whole science of astronomy had been profoundly changed by the invention of the telescope. Instruments called "perspective glasses" had been used in England before 1580 for viewing distant terrestrial objects, and both John Dee and Thomas Digges were known to be expert in their construction and use, though there is no evidence that they used them for astronomical observations. But their friend Thomas Harriot, the Wizard Earl, is known to have made astronomical observations in the winter of 1609–10 with a small "telescope," which may have been a perspective glass.

Other than these perspective glasses, one of the earliest telescopes seems to have appeared in 1604, when a Dutch optician named Zacharias Janssen constructed one from a specimen belonging to an unknown Italian, after which he sold some of them at fairs in northern Europe. Hearing of the telescope, Galileo built one in his workshop in 1609 and then offered it to the Doge of Venice for use in war and navigation. After improving on his original design, he began using his telescope to observe the heavens, and in March 1610 he published his discoveries in a little book called *Siderius Nuncius* (The Starry Messenger).

The book begins with his observations of the moon, which he found to look very much like the earth, with mountains, valleys, and what he thought were seas. Seen in the telescope, the planets were pale illuminated disks, whereas the stars remained brilliant points of light. The Milky Way proved to consist of numerous stars, not a nebula reflecting the light of the sun, as some had thought, nor an atmospheric phenomenon, as Aristotle had concluded. He counted more than ninety stars in Orion's belt, where only nine are visible to the naked eye. He discovered four moons orbiting around Jupiter, a solar system in miniature, which he used as an additional argument in favor of the Copernican theory. He called the Jovian moons the "Medicean Stars" in honor of Cosimo de' Medici, the Grand Duke of Tuscany. Cosimo responded by making Galileo his court philosopher and appointing him to the chair of mathematics at the University of Pisa. Galileo had no obligation to teach at the University of Pisa or even to reside in the city, and so after his appointment, in September 1610, he departed to take up residence in Florence.

Galileo sent a copy of the *Siderius Nuncius* to Kepler, who received it on 8 April 1610. During the next eleven days Kepler composed his response in a little work called *Dissertatio cum Nuncio Sidereal* (Answer to the Starry Messenger), in which he expressed his enthusiastic approval of Galileo's discoveries and re-minded readers of his own work on optical astronomy, as well as speculating on the possibility of inhabitants on the moon and arguing against an infinite universe.

Kepler borrowed a telescope from the elector Ernest of Cologne at the end of August 1610, and for the next ten days he used it to observe the heavens, particularly Jupiter and its moons. His excitement over the possibilities of the new instrument was such that he spent the next two months making an exhaustive study of the passage of light through lenses, which he published later in 1610 under the title *Dioptrice,* which became one of the foundation stones of the new science of optics.

The death of Rudolph II in early 1612 forced Kepler to leave Prague and take up the post of district mathematician at Linz, where he remained for the next fourteen years. One of his official duties was a study of chronology, part of a program of calendar reform instituted by Archduke Ferdinand II, son of the late emperor Rudolph. As a result of his studies he established that Christ was born in what in the modern calendar would be 5 B.C.

During that period when Kepler lived in Linz he continued his calculations on the *Rudolfine Tables* and published two other major works, the first of which was the *Harmonice Mundi* (Harmony of the World), which appeared in 1619. The title of this work was inspired by a Greek manuscript of Ptolemy's treatise on musical theory, the *Harmonica,* which Kepler acquired in 1607 and used in his analysis of music, geometry, astronomy, and astrology. The most important part of the *Harmonice Mundi* is the relationship now known as Kepler's third law of planetary motion, which he discovered on 15 May 1618 and presented in Book V. The law states that for each of the planets the square of the period of its orbital motion is proportional to the cube of its distance from the sun (or, strictly speaking, the semimajor axis of its elliptical orbit).

There had been speculations about the relation between the periods of planetary orbits and their radii since the times of Pythagoras, Plato, and Aristotle, and Kepler was terribly excited that he had at last, following in the footsteps of Ptolemy, found the mathematical law "necessary for the contemplation of celestial harmonies." He wrote of his pleasure, "That the same thought about the harmonic formulation had turned up in the minds of two men (though lying so far apart in time) who had devoted themselves entirely to contemplating nature ... I feel carried away and possessed by an unutterable rapture over the divine spectacle of the heavenly harmony."

Kepler dedicated the *Harmonice* to James I of England. The king responded by sending his ambassador Sir Henry Wooton with an invitation for Kepler to take up residence in England. But after considering the offer for a while Kepler eventually decided against it.

The English poet John Donne was familiar with the work of Copernicus and Kepler, probably through Thomas Harriot. Donne had in 1611 said to the Copernicans that "those opinions of yours may very well be true ... creeping into every man's mind." That same year Donne lamented the passing of the old cosmology in "An Anatomy of the World":

> *And new Philosophy cals all in doubt,*
> *The Element of fire is quite put out;*
> *The Sun is lost, and th'earth, and no man's wit*
> *Can well direct him, where to look for it.*

Kepler's second major work at Linz was his *Epitome Astronomiae Coper-nicanae* (Compendium of Copernican Astronomy), published in 1621. In the first three of the seven books of the *Epitome* Kepler refutes the traditional arguments against the motions of the earth, going much further than Copernicus and using principles that Galileo would later give in greater detail. His three laws of planetary motion are explained in great detail in Book IV, along with his lunar theory. The last three books treat practical problems involving his first two laws of planetary motion as well as his theories of lunar and solar motion and the precession of the equinoxes.

In 1626 Kepler was forced to leave Linz, which had undergone a two-month siege during a peasant uprising, and move to Ulm, where he published the *Rudolfine Tables* in September 1627, dedicating them to Archduke Ferdinand II. The new tables were far more accurate than any in the past, and they remained in use for more than a century. Kepler used his tables to predict that Mercury and Venus would make transits across the disk of the sun in 1631. The transit of Venus was not observed in Europe because it took place at night. The transit of Mercury was observed by Pierre Gassendi in Paris on 7 November 1631, representing a triumph for Kepler's astronomy, for his prediction was in error by only 10 minutes of arc as compared to 5 degrees for tables based on Ptolemy's model.

But Kepler did not live to see his theories vindicated, for he passed away on 15 November 1630. His tombstone, now lost, was engraved with an epitaph that he had written himself:

> *I used to measure the heavens,*
> *Now I measure the shadow of the earth.*

> *Although my soul was from heaven,*
> *The shadow of my body lies here.*

Meanwhile, Galileo had been active in advancing the cause of Copernicanism against the accepted cosmology of Aristotle, which in its reinterpretation by Saint Thomas Aquinas formed part of the philosophical basis for Roman Catholic theology. At the beginning of March 1616 the Holy Office of the Inquisition in Rome placed the works of Copernicus and all other writings that supported it, including those of Kepler, on the Index, the list of books that Catholics were forbidden to read. The decree held that believing the sun to be the immovable center of the world is "foolish and absurd, philosophically false and formally heretical." Pope Paul V instructed Cardinal Bellarmine to censure Galileo, admonishing him not to hold or defend Copernican doctrines any longer. On March 3 Bellarmine reported that Galileo had acquiesced to the pope's warning, and that ended the matter for the time being.

After his censure Galileo returned to his villa at Arcetri, outside Florence, where for the next seven years he remained silent.

How the Heavens Go

By William H. Cropper

The Tale of the Tower

Legend has it that a young, ambitious, and at that moment frustrated mathematics professor climbed to the top of the bell tower in Pisa one day, perhaps in 1591, with a bag of ebony and lead balls. He had advertised to the university community at Pisa that he intended to disprove by experiment a doctrine originated by Aristotle almost two thousand years earlier: that objects fall at a rate proportional to their weight; a ten-pound ball would fall ten times faster than a one-pound ball. With a flourish the young professor signaled to the crowd of amused students and disapproving philosophy professors below, selected balls of the same material but with much different weights, and dropped them. Without air resistance (that is, in a vacuum), two balls of different weights (and made of any material) would have reached the ground at the same time. That did not happen in Pisa on that day in 1591, but Aristotle's ancient principle was clearly violated anyway, and that, the young professor told his audience, was the lesson. The students cheered, and the philosophy professors were skeptical.

The hero of this tale was Galileo Galilei. He did not actually conduct that "experiment" from the Tower of Pisa, but had he done so it would have been entirely in character. Throughout his life. Galileo had little regard for authority, and one of his perennial targets was Aristotle, the ultimate authority for university philosophy faculties at the time. Galileo's personal style was confrontational, witty, ironic, and often sarcastic. His intellectual style, as the Tower story instructs, was to build his theories with an ultimate appeal to observations.

The philosophers of Pisa were not impressed with either Galileo or his methods, and would not have been any more sympathetic even if they had witnessed the Tower experiment. To no one's surprise, Galileo's contract at the University of Pisa was not renewed.

Padua

But Galileo knew how to get what he wanted. He had obtained the Pisa post with the help of the Marquis Guidobaldo del Monte, an influential nobleman and competent mathematician. Galileo now aimed for the recently vacated chair of mathematics at the University of Padua, and his chief backer in Padua was Gian-vincenzio Pinelli, a powerful influence in the cultural and intellectual life of Padua. Galileo followed Pinelli's advice, charmed the examiners, and won the approval of the Venetian senate (Padua was located in the Republic of Venice, about twenty miles west of the city of Venice). His inaugural lecture was a sensation.

Padua offered a far more congenial atmosphere for Galileo's talents and lifestyle than the intellectual backwater he had found in Pisa. In the nearby city of Venice, he found recreation and more—aristocratic friends. Galileo's favorite debating partner among these was Gianfrancesco Sagredo, a wealthy nobleman with an eccentric manner Galileo could appreciate. With his wit and flair for polemics, Galileo was soon at home in the city's salons. He took a mistress, Marina Gamba, described by one of Galileo's biographers, James Reston, Jr., as "hot-tempered, strapping, lusty and probably illiterate." Galileo and Marina had three children: two daughters, Virginia and Livia, and a son, Vincenzo. In later life, when tragedy loomed, Galileo found great comfort in the company of his elder daughter, Virginia.

During his eighteen years in Padua (1592–1610), Galileo made some of his most important discoveries in mechanics and astronomy. From careful observations, he formulated the "times-squared" law, which states that the vertical distance covered by an object in free fall or along an inclined plane is proportional to the square of the time of the fall. (In modern notation, the equation for free gt^2 fall is expressed $s = gt^2/2$, with s and t the vertical distance and time of the fall (and g the acceleration of gravity.) He defined the laws of projected motion with a controlled version of the Tower experiment in which a ball rolled down an inclined plane on a table, then left the table horizontally or obliquely and dropped to the floor. Galileo found that he could make calculations that agreed approximately with his experiments by resolving projected motion into two components, one horizontal and the other vertical. The horizontal component was determined by the speed of the ball when it left the table, and was "conserved"—that is, it did not subsequently change. The vertical component, due to the ball's weight, followed the times-squared rule.

For many years, Galileo had been fascinated by the simplicity and regularity of pendulum motion. He was most impressed by the constancy of the pendulum's "period," that is, the time the pendulum takes to complete its back-and-forth cycle. If the pendulum's swing is less than about 30°, its period is, to a good approximation, dependent only on its length. (Another Galileo legend pictures him as a nineteen-year-old boy in church, paying little attention to the service, and timing with his pulse the swings of an oil lamp suspended on a wire from a high ceiling.) In Padua, Galileo confirmed the constant-period rule with experiments, and then uncovered some of the pendulum's more subtle secrets.

In 1609, word came to Venice that spectacle makers In Holland had invented an optical device—soon to be called a telescope—that brought distant objects much closer. Galileo immediately saw a shining opportunity. If he could build a prototype and demonstrate it to the Venetian authorities before Dutch entrepreneurs arrived on the scene, unprecedented rewards would follow. He knew enough about optics to guess that the Dutch design was a combination of a convex and a concave lens, and he and his instrument maker had the exceptional skill needed to grind the lenses. In twenty-four hours, according to Galileo's own account, he had a telescope of better quality than any produced by the Dutch artisans. Galileo could have demanded, and no doubt received, a large sum for his invention. But fame and influence meant more to him than money. In an elaborate ceremony, he gave an eight-power telescope to Niccolo Contarini, the doge of Venice. Reston, in *Galileo,* paints this picture of the presentation of the telescope: "a celebration of Venetian genius, complete with brocaded advance men, distinguished heralds and secret operatives. Suddenly, the tube represented the flowering of Paduan learning." Galileo was granted a large bonus, his salary was doubled, and he was reappointed to his faculty position for life.

Then Galileo turned his telescope to the sky, and made some momentous, and as it turned out fateful, discoveries. During the next several years, he observed the mountainous surface of the Moon, four of the moons of Jupiter, the phases of Venus, the rings of Saturn (not quite resolved by his telescope), and sunspots. In 1610, he published his observations in *The Starry Messenger*, which was an immediate sensation, not only in Italy but throughout Europe.

But Galileo wanted more. He now contrived to return to Tuscany and Florence, where he had spent most of his early life. The grand duke of Tuscany was the young Cosimo de Medici, recently one of Galileo's pupils. To further his cause, Galileo dedicated *The Starry Messenger* to the grand duke and named the four moons of Jupiter the Medicean satellites. The flattery had its intended effect. Galileo soon accepted an astonishing offer from Florence: a salary equivalent to that of the highest-paid court official, no lecturing duties—in fact, no duties of any kind—and the title of chief mathematician and philosopher for the grand duke of Tuscany. In Venice and Padua, Galileo left behind envy and bitterness.

Florence and Rome

Again the gregarious and witty Galileo found intellectual companions among the nobility. Most valued now was his friendship with the young, talented, and skeptical Filippo Salviati. Galileo and his students were regular visitors at Salviati's beautiful villa fifteen miles from Florence. But even in this idyll Galileo was restless. He had one more world to conquer: Rome—that is, the Church. In 1611, Galileo proposed to the grand duke's secretary of state an official visit to Rome in which he would demonstrate his telescopes and impress the Vatican with the importance of his astronomical discoveries.

This campaign had its perils. Among Galileo's discoveries was disturbing evidence against the Church's doctrine that Earth was the center of the universe. The Greek astronomer and mathematician Ptolemy had advocated this cosmology in the second

century, and it had long been Church dogma. Galileo could see in his observations evidence that the motion of Jupiter's moons centered on Jupiter, and, more troubling, in the phases of Venus that the motion of that planet centered on the Sun. In the sixteenth century, the Polish astronomer Nicolaus Copernicus had proposed a cosmology that placed the Sun at the center of the universe. By 1611, when he journeyed to Rome, Galileo had become largely converted to Copernicanism. Holy Scripture also regarded the Moon and the Sun as quintessentially perfect bodies; Galileo's telescope had revealed mountains and valleys on the Moon and spots on the Sun.

But in 1611 the conflict between telescope and Church was temporarily submerged, and Galileo's stay was largely a success. He met with the autocratic Pope Paul V and received his blessing and support. At that time and later, the intellectual power behind the papal throne was Cardinal Robert Bellarmine. It was his task to evaluate Galileo's claims and promulgate an official position. He, in turn, requested an opinion from the astronomers and mathematicians at the Jesuit Collegio Romano, who reported doubts that the telescope really revealed mountains on the Moon, but more importantly, trusted the telescope's evidence for the phases of Venus and the motion of Jupiter's moons.

Galileo found a new aristocratic benefactor in Rome. He was Prince Frederico Cesi, the founder and leader of the "Academy of Lynxes," a secret society whose members were "philosophers who are eager for real knowledge, and who will give themselves to the study of nature, and especially to mathematics." The members were young, radical, and, true to the lynx metaphor, sharp-eyed and ruthless in their treatment of enemies. Galileo was guest of honor at an extravagant banquet put on by Cesi, and shortly thereafter was elected as one of the Lynxes.

Galileo gained many influential friends in Rome and Florence—and, inevitably, a few dedicated enemies. Chief among those in Florence was Ludovico della Colombe, who became the self-appointed leader of Galileo's critics. *Colombe* means "dove" in Italian. Galileo expressed his contempt by calling Colombe and company the "Pigeon League."

Late in 1611, Colombe, whose credentials were unimpressive, went on the attack and challenged Galileo to an intellectual duel: a public debate on the theory of floating bodies, especially ice. A formal challenge was delivered to Galileo by a Pisan professor, and Galileo cheerfully responded, "Ever ready to learn from anyone, I should take it as a favor to converse with this friend of yours and reason about the subject." The site of the debate was the Pitti Palace. In the audience were two cardinals, Grand Duke Cosimo, and Grand Duchess Christine, Cosimo's mother. One of the cardinals was Maffeo Barberini, who would later become Pope Urban VIII and play a major role in the final act of the Galileo drama.

In the debate, Galileo took the view that ice and other solid bodies float because they are lighter than the liquid in which they are immersed. Colombe held to the Aristotelian position that a thin, flat piece of ice floats in liquid water because of its peculiar shape. As usual, Galileo built his argument with demonstrations. He won the audience, including Cardinal Barberini, when he showed that pieces of ebony, even in very thin shapes, always sank in water, while a block of ice remained on the surface.

The Gathering Storm

The day after his victory in the debate, Galileo became seriously ill, and he retreated to Salviati's villa to recuperate. When he had the strength, Galileo summarized in a treatise his views on floating bodies, and, with Salviati, returned to the study of sunspots. They mapped the motion of large spots as the spots traveled across the sun's surface near the equator from west to east.

Then, in the spring of 1612, word came that Galileo and Salviati had a competitor. He called himself Apelles. (He was later identified as Father Christopher Scheiner, a Jesuit professor of mathematics in Bavaria.) To Galileo's dismay, Apelles claimed that his observations of sunspots were the first, and explained the spots as images of stars passing in front of the sun. Not only was the interloper encroaching on Galileo's priority claim, but he was also broadcasting a false interpretation of the spots. Galileo always had an inclination to paranoia, and it now had the upper hand. He sent a series of bold letters to Apelles through an intermediary, and agreed with Cesi that the letters should be published in Rome by the Academy of Lynxes. In these letters Galileo asserted for the first time his adherence to the Copernican cosmology. As evidence he recalled his observations of the planets: "I tell you that [Saturn] also, no less than the horned Venus agrees admirably with the great Copernican system. Favorable winds are now blowing on that system. Little reason remains to fear crosswinds and shadows on so bright a guide."

Galileo soon had another occasion to proclaim his belief in Copernicanism. One of his disciples, Benedetto Castelli, occupied Galileo's former post, the chair of mathematics at Pisa. In a letter to Galileo, Castelli wrote that recently he had had a disturbing interview with the pious Grand Duchess Christine. "Her Ladyship began to argue against me by means of the Holy Scripture," Castelli wrote. Her particular concern was a passage from the Book of Joshua that tells of God commanding the Sun to stand still so Joshua's retreating enemies could not escape into the night. Did this not support the doctrine that the Sun moved around Earth and deny the Copernican claim that Earth moved and the Sun was stationary?

Galileo sensed danger. The grand duchess was powerful, and he feared that he was losing her support. For the first time he openly brought his Copernican views to bear on theological issues. First he wrote a letter to Castelli. It was sometimes a mistake, he wrote, to take the words of the Bible literally. The Bible had to be interpreted in such a way that there was no contradiction with direct observations: "The task of wise interpreters is to find true meanings of scriptural passages that will agree with the evidence of sensory experience." He argued that God could have helped Joshua just as easily under the Copernican cosmology as under the Ptolemaic.

The letter to Castelli, which was circulated and eventually published, brought no critical response for more than a year. In the meantime, Galileo took more drastic measures. He expanded the letter, emphasizing the primacy of observations over doctrine when the two were in conflict, and addressed it directly to Grand Duchess Christine. "The primary purpose of the Holy Writ is to worship God and save souls," he wrote. But "in disputes about natural phenomena, one must not begin with the authority of scriptural passages, but with sensory experience and necessary demonstrations." He recalled that Cardinal Cesare Bar-onius had once said, "The Bible tells us how to go to Heaven, not how the heavens go."

The first attack on Galileo from the pulpit came from a young Dominican priest named Tommaso Caccini, who delivered a furious sermon centering on the miracle of Joshua, and the futility of understanding such grand events without faith in established doctrine. This was a turning point in the Galileo story. As Reston puts it: "Italy's most famous scientist, philosopher to the Grand Duke of Tuscany, intimate of powerful cardinals in Rome, stood accused publicly of heresy from an important pulpit, by a vigilante of the faith." Caccini and Father Niccolo

Lorini, another Dominican priest, now took the Galileo matter to the Roman Inquisition, presenting as evidence for heresy the letter to Castelli.

Galileo could not ignore these events. He would have to travel to Rome and face the inquisitors, probably influenced by Cardinal Bellarmine, who had, four years earlier, reported favorably on Galileo's astronomical observations. But once again Galileo was incapacitated for months by illness. Finally, in late 1615 he set out for Rome.

As preparation for the inquisitors, a Vatican commission had examined the Copernican doctrine and found that its propositions, such as placing the Sun at the center of the universe, were "foolish and absurd and formally heretical." On February 25, 1616, the Inquisition met and received instructions from Pope Paul to direct Galileo not to teach or defend or discuss Copernican doctrine. Disobedience would bring imprisonment.

In the morning of the next day, Bellarmine and an inquisitor presented this injunction to Galileo orally. Galileo accepted the decision without protest and waited for the formal edict from the Vatican. That edict, when it came a few weeks later, was strangely at odds with the judgment delivered earlier by Bellarmine. It did not mention Galileo or his publications at all, but instead issued a general restriction on Copernicanism: "It has come to the knowledge of the Sacred Congregation that the false Pythagorean doctrine, namely, concerning the movement of the Earth and immobility of the Sun, taught by Nicolaus Copernicus, and altogether contrary to the Holy Scripture, is already spread about and received by many persons. Therefore, lest any opinion of this kind insinuate itself to the detriment of Catholic truth, the Congregation has decreed that the works of Nicolaus Copernicus be suspended until they are corrected."

Galileo, always an optimist, was encouraged by this turn of events. Despite Bellarmine's strict injunction, Galileo had escaped personal censure, and when the "corrections" to Copernicus were spelled out they were minor. Galileo remained in Rome for three months, and found occasions to be as outspoken as ever. Finally, the Tuscan secretary of state advised him not to "tease the sleeping dog further," adding that there were "rumors we do not like."

Comets, a Manifesto, and a Dialogue

In Florence again, Galileo was ill and depressed during much of 1617 and 1618. He did not have the strength to comment when three comets appeared in the night sky during the last four months of 1618. He was stirred to action, however, when Father Horatio Grassi, a mathematics professor at the Collegio Romano and a gifted scholar, published a book in which he argued that the comets provided fresh evidence against the Copernican cosmology. At first Galileo was too weak to respond himself, so he assigned the task to one of his disciples, Mario Gui-ducci, a lawyer and graduate of the Collegio Romano. A pamphlet, *Discourse on Comets,* was published under Guiducci's name, although the arguments were clearly those of Galileo.

This brought a worthy response from Grassi, and in 1621 and 1622 Galileo was sufficiently provoked and healthy to publish his eloquent manifesto, *The Assayer.* Here Galileo proclaimed, "Philosophy is written in this grand book the universe, which stands continually open to our gaze. But the book cannot be understood unless one first learns to comprehend the language and to read the alphabet in which it is composed. It is written in the language of mathematics, and its characters are triangles, circles and other geometric figures, without which it is humanly impossible to understand a single word of it; without these, one wanders about in a dark labyrinth."

The Assayer received Vatican approval, and Cardinal Barberini, who had supported Galileo in his debate with del la Colombe, wrote in a friendly and reassuring letter, "We are ready to serve you always." As it turned out, Barberini's good wishes could hardly have been more opportune. In 1623, he was elected pope and took the name Urban VIII.

After recovering from a winter of poor health, Galileo again traveled to Rome in the spring of 1624. He now went bearing microscopes. The original microscope design, like that of the telescope, had come from Holland, but Galileo had greatly improved the instrument for scientific uses. Particularly astonishing to the Roman cognoscenti were magnified images of insects.

Shortly after his arrival in Rome, Galileo had an audience with the recently elected Urban VIII.

Expecting the former Cardinal Barberini again to promise support, Galileo found to his dismay a different persona. The new pope was autocratic, given to nepotism, long-winded, and obsessed with military campaigns. Nevertheless, Galileo left Rome convinced that he still had a clear path. In a letter to Cesi he wrote, "On the question of Copernicus His Holiness said that the Holy Church had not condemned, nor would condemn his opinions as heretical, but only rash. So long as it is not demonstrated as true, it need not be feared."

Galileo's strategy now was to present his arguments hypothetically, without claiming absolute truth. His literary device was the dialogue. He created three characters who would debate the merits of the Copernican and Aristotelian systems, but ostensibly the debate would have no resolution. Two of the characters were named in affectionate memory of his Florentine and Venetian friends, Gian-francesco Sagredo and Filippo Salviati, who had both died. In the dialogue Salviati speaks for Galileo, and Sagredo as an intelligent layman. The third character is an Aristotelian, and in Galileo's hands earns his name, Simplicio.

The dialogue, with the full title *Dialogue Concerning the Two Chief World Systems,* occupied Galileo intermittently for five years, between 1624 and 1629. Finally, in 1629, it was ready for publication and Galileo traveled to Rome to expedite approval by the Church. He met with Urban and came away convinced that there were no serious obstacles.

Then came some alarming developments. First, Cesi died. Galileo had hoped to have his *Dialogue* published by Cesi's Academy of Lynxes, and had counted on Cesi as his surrogate in Rome. Now with the death of Cesi, Galileo did not know where to turn. Even more alarming was an urgent letter from Castelli advising him to publish the *Dialogue* as soon as possible in Florence. Galileo agreed, partly because at the time Rome and Florence were isolated by an epidemic of bubonic plague. In the midst of the plague, Galileo found a printer in Florence, and the printing was accomplished. But approval by the Church was not granted for two years, and when the *Dialogue* was finally published it contained a preface and conclusion written by the Roman Inquisitor. At first, the book found a sympathetic audience. Readers were impressed by Galileo's accomplished use of the dialogue form, and they found the dramatis personae, even the satirical Simplicio, entertaining.

In August 1632, Galileo's publisher received an order from the Inquisition to cease printing and selling the book. Behind this sudden move was the wrath of

Urban, who was not amused by the clever arguments of Salviati and Sagredo, and the feeble responses of Simplicio. He even detected in the words of Simplicio some of his own views. Urban appointed a committee headed by his nephew, Cardinal Francesco Barberini, to review the book. In September, the committee reported to Urban and the matter was handed over to the Inquisition.

Trial

After many delays—Galileo was once again seriously ill, and the plague had returned—Galileo arrived in Rome in February 1633 to defend himself before the Inquisition. The trial began on April 12. The inquisitors focused their attention on the injunction Bellarmine had issued to Galileo in 1616. Francesco Niccolini, the Tuscan ambassador to Rome, explained it this way to his office in Florence: "The main difficulty consists in this: these gentlemen [the inquisitors] maintain that in 1616 he [Galileo] was commanded neither to discuss the question of the earth's motion nor to converse about it. He says, to the contrary, that these were not the terms of the injunction, which were that that doctrine was not to be held or defended. He considers that he has the means of justifying himself since it does not appear at all from his book that he holds or defends the doctrine … or that he regards it as a settled question." Galileo offered in evidence a

letter from Bellarmine, which bolstered his claim, that the inquisitors' strict interpretation of the injunction was not valid.

Historians have argued about the weight of evidence on both sides, and on a strictly legal basis, concluded that Galileo had the stronger case. (Among other things, the 1616 injunction had never been signed or witnessed.) But for the inquisitors, acquittal was not an option. They offered what appeared to be a reasonable settlement: Galileo would admit wrongdoing, submit a defense, and receive a light sentence. Galileo agreed and complied. But when the sentence came on June 22 it was far harsher than anything he had expected: his book was to be placed on the Index of Prohibited Books, and he was condemned to life imprisonment.

Last Act

Galileo's friends always vastly outnumbered his enemies. Now that he had been defeated by his enemies, his friends came forward to repair the damage. Ambassador Niccolini managed to have the sentence commuted to custody under the Archbishop Ascanio Piccolomini of Siena. Galileo's "prison" was the archbishop's palace in Siena, frequented by poets, scientists, and musicians, all of whom arrived to honor Galileo. Gradually his mind returned to the problems of science, to topics that were safe from theological entanglements. He planned a dialogue on "two new sciences," which would summarize his work on natural motion (one science) and also address problems related to the strengths of materials (the other science). His three interlocutors would again be named Salviati, Sagredo, and Simplicio, but now they would represent three ages of the author: Salviati, the wise Galileo in old age; Sagredo, the Galileo of the middle years in Padua; and Simplicio, a youthful Galileo.

But Galileo could not remain in Siena. Letters from his daughter Virginia, now Sister Maria Celeste in the convent of St. Matthew in the town of Arcetri, near Florence, stirred deep memories. Earlier he had taken a villa in Arcetri to be near Virginia and his other daughter, Livia, also a sister at the convent. He now appealed to the pope for permission to return to Arcetri. Eventually the request was granted, but only after word had come that Maria Celeste was seriously ill, and more important, after the pope's agents had reported that the heretic's comfortable "punishment" in Siena did not fit the crime. The pope's edict directed that Galileo return to his villa and remain guarded there under house arrest.

Galileo took up residence in Arcetri in late 1633, and for several months attended Virginia in her illness. She did not recover, and in the spring of 1634, she died. For Galileo this was almost the final blow. But once again work was his restorative. For three years he concentrated on his *Discourses on Two New Sciences,* That work, his final masterpiece, was completed in 1637, and in 1638 it was published (in Holland, after the manuscript was smuggled out of Italy). By this time Galileo had gone blind. Only grudgingly did Urban permit Galileo to travel the short distance to Florence for medical treatment.

But after all he had endured, Galileo never lost his faith. "Galileo's own conscience was clear, both as Catholic and as scientist," Stillman Drake, a contemporary science historian, writes. "On one occasion he wrote, almost in despair, that he felt like burning all his work in science; but he never so much as thought of turning his back on his faith. The Church turned its back on Galileo, and has suffered not a little for having done so; Galileo blamed only some wrongheaded individuals in the Church for that."

Methods

Galileo's mathematical equipment was primitive. Most of the mathematical methods we take for granted today either had not been discovered or had not come into reliable use in Galileo's time. He did not employ algebraic symbols or equations, or, except

for tangents, the concepts of trigonometry. His numbers were always expressed as positive integers, never as decimals. Calculus, discovered later by Newton and Gottfried Leibniz, was not available. To make calculations he relied on ratios and proportionalities, as defined in Euclid's *Elements.* His reasoning was mostly geometric, also learned from Euclid.

Galileo's mathematical style is evident in his many theorems on uniform and accelerated motion; here a few are presented and then "modernized" through translation into the language of algebra. The first theorem concerns uniform motion:

> If a moving particle, carried uniformly at constant speed, traverses two distances, the time intervals required are to each other in the ratio of these distances.

For us (but not for Galileo) this theorem is based on the algebraic equation $s = Vt$, in which s represents distance, V speed, and t time. This is a familiar calculation. For example, if you travel for three hours ($t = 3$ hours) at sixty miles per hour ($V = 60$ miles per hour), the distance you have covered is 180 miles ($s = 3 \times 60 = 180$ miles). In Galileo's theorem, we calculate two distances, call them s_1 and s_2, for two times. t_1 and t_2, at the same speed, $V.$ The two calculations are

$$s_1 = Vt_1 \text{ and } s_2 = Vt_2$$

Dividing the two sides of these equations into each other, we get the ratio of Galileo's theorem,

$$t_1 / t_2 = s_1 / s_2$$

Here is a more complicated theorem, which does not require that the two speeds be equal:

> If two particles are moved at a uniform rate, but with unequal speeds, through unequal distances, then the ratio of time intervals

occupied will be the product of the ratio of the distances by the inverse ratio of the speeds.

In this theorem, there are two different speeds, V_1 and V_2, involved, and the two equations are

$$s_1 = v_1 t_1 \text{ and } s_2 = v_2 t_2$$

Dividing both sides of the equations into each other again, we have

$$s_1 / s_2 = v_1 / v_2 \, t_1 / t_2$$

To finish the proof of the theorem, we multiply both sides of this equation by v_2 / v_1 and obtain

$$t_1 / t_2 = s_1 / s_2 \, v_2 / v_1$$

On the right side now is a product of the direct ratio of the distances s_1 / s_2 and the inverse ratio of the speeds v_2 / v_1, as required by the theorem.

These theorems assume that any speed v is constant; that is, the motion is not accelerated. One of Galileo's most important contributions was his treatment of uniformly accelerated motion, both in free fall and down inclined planes. "Uniformly" here means that the speed changes by equal amounts in equal time intervals. If the uniform acceleration is represented by a, the change in the speed V in time t is calculated with the equation $V = at$. For example, if you accelerate your car at the uniform rate $a = 5$ miles per hour per second for $t = 10$ seconds, your final speed will be $V = 5 \times 10 = 50$ miles per hour. A second equation, $s = at^2/2$, calculates s, the distance covered in time t under the uniform acceleration a. This equation is not so familiar as the others mentioned. It is most easily justified with the methods of calculus, as will be demonstrated in the next chapter.

The motion of a ball of any weight dropping in free fall is accelerated in the vertical direction, that is, perpendicular to Earth's surface, at a rate that is conventionally represented by the symbol g, and is nearly the same anywhere on Earth. For the case of free fall, with $a = g$, the last two equations mentioned are $V = gt$, for the speed attained in free fall in the time t, and $s = gt^2/2$ for the corresponding distance covered.

Galileo did not use the equation $s = gt^2/2$, but he did discover through experimental observations the times-squared (t^2) part of it. His conclusion is expressed in the theorem,

The spaces described by a body falling from rest with a uniformly accelerated motion are to each other as the squares of the time intervals employed in traversing these distances.

Our modernized proof of the theorem begins by writing the free-fall equation twice,

$$s_1 = gt_1^2/2 \text{ and } s_2 = gt_2^2/2$$

and combining these two equations to obtain

$$s_1/s_2 = t_1^2/t_2^2$$

In addition to his separate studies of uniform and accelerated motion, Galileo also treated a composite of the two in projectile motion. He proved that the trajectory followed by a projectile is parabolic. Using a complicated geometric method, he developed a formula for calculating the dimensions of the parabola followed by a projectile (for example, a cannonball) launched upward at any angle of elevation. The formula is cumbersome compared to the trigonometric method we use today for such calculations, but no less accurate. Galileo demonstrated the use of his method by calculating with remarkable precision a detailed table of parabola dimensions for angles of elevation from 1° to 89°.

In contrast to his mathematical methods, derived mainly from Euclid, Galileo's experimental methods seem to us more modern. He devised a system of units that parallels our own and that served him well in his experiments on pendulum motion. His measure of distance, which he called a *punto,* was equivalent to 0.094 centimeter. This was the distance between the finest divisions on a brass rule. For measurements of time he collected and weighed water flowing from a container at a constant rate of about three fluid ounces per second. He recorded weights of water in grains (1 ounce = 480 grains), and defined his time unit, called a *tempo,* to be the time for 16 grains of water to flow, which was equivalent to 1/92 second. These units were small enough so Galileo's measurements of distance and time always resulted in large numbers. That was a necessity because decimal numbers were not part of his mathematical equipment; the only way he could add significant digits in his calculations was to make the numbers larger.

Legacy

Galileo took the metaphysics out of physics, and so begins the story that will unfold in the remaining chapters of this book. As Stephen Hawking writes, "Galileo, perhaps more than any single person, was responsible for the birth of modern science. … Galileo was one of the first to argue that man could hope to understand how the world works, and, moreover, that he could do this by observing the real world." No practicing physicist, or any other scientist for that matter, can do his or her work without following this Galilean advice.

I have already mentioned many of Galileo's specific achievements. His work in mechanics is worth sketching again, however, because it paved the way for his greatest successor. (Galileo died in January 1642. On Christmas Day of that same year, Isaac Newton was born.) Galileo's mechanics is largely concerned with bodies moving at constant velocity or under constant acceleration, usually that of gravity. In our view, the theorems that define his mechanics are based

on the equations $V = gt$ and $s = gt^2/2$, but Galileo did not write these, or any other, algebraic equations; for his numerical calculations he invoked ratios and proportionality. He saw that projectile motion was a resultant of a vertical component governed by the acceleration of gravity and a constant horizontal component given to the projectile when it was launched. This was an early recognition that physical quantities with direction, now called "vectors," could be resolved into rectangular components.

I have mentioned, but not emphasized, another building block of Galileo's mechanics, what is now called the "inertia principle." In one version, Galileo put it this way: "Imagine any particle projected along a horizontal plane without friction; then we know… that this particle will move along this plane with a motion which is uniform and perpetual, provided the plane has no limits." This statement reflects Galileo's genius for abstracting a fundamental idealization from real behavior. If you give a real ball a push on a real horizontal plane, it will not continue its motion perpetually, because neither the ball nor the plane is perfectly smooth, and sooner or later the ball will stop because of frictional effects. Galileo neglected all the complexities of friction and obtained a useful postulate for his mechanics. He then applied the postulate in his treatment of projectile motion. When a projectile is launched, its horizontal component of motion is constant in the absence of air resistance, and remains that way, while the vertical component is influenced by gravity.

Galileo's mechanics did not include definitions of the concepts of force or energy, both of which became important in the mechanics of his successors. He had no way to measure these quantities, so he included them only in a qualitative way. Galileo's science of motion contains most of the ingredients of what we now call "kinematics." It shows us how motion occurs without defining the forces that control the motion. With the forces included, as in Newton's mechanics, kinematics becomes "dynamics."

All of these specific Galilean contributions to the science of mechanics were essential to Newton and his successors. But transcending all his other contributions was Galileo's unrelenting insistence that the success or failure of a scientific theory depends on observations and measurements. Still man Drake leaves us with this trenchant synopsis of Galileo's scientific contributions: "When Galileo was born, two thousand years of physics had not resulted in even rough measurements of actual motions. It is a striking fact that the history of each science shows continuity back to its first use of measurement, before which it exhibits no ancestry but metaphysics. That explains why Galileo's science was stoutly opposed by nearly every philosopher of his time, he having made it as nearly free from metaphysics as he could. That was achieved by measurements, made as precisely as possible with means available to Galileo or that he managed to devise."

The Scientists

By John Gribbin

Isaac Newton came, on his father's side, from a farming family that had just started to do well for themselves in a material way, but lacked any pretensions to intellectual achievement. His grandfather, Robert Newton, had been born some time around 1570 and had inherited farmland at Woolsthorpe, in Lincolnshire. He prospered from his farming so much that he was able to purchase the manor of Woolsthorpe in 1623, gaining the title of Lord of the Manor. Though not as impressive as it sounds to modern ears, this was a distinct step up the social ladder for the Newton family, and probably an important factor in enabling Robert's son Isaac (born in 1606) to marry Hannah Ayscough, the daughter of James Ayscough, described in contemporary accounts as 'a gentleman'. The betrothal took place in 1639. Robert made Isaac the heir to all his property, including the lordship of the manor, and Hannah brought to the union as her dowry property worth £50 per year. Neither Robert Newton nor his son Isaac ever learned to read or write, but Hannah's brother William was a Cambridge graduate, a clergyman who enjoyed the living at the nearby village of Burton Coggles. The marriage between Isaac and Hannah took place in 1642, six months after the death of Robert Newton; six months after the wedding, Isaac also died, leaving Hannah pregnant with a baby who was born on Christmas Day and christened Isaac after his late father.

Many popular accounts note the coincidence that 'the' Isaac Newton was born in the same year, 1642, that Galileo died. But the coincidence rests on a cheat-using dates from two different calendar systems. Galileo died on 8 January 1642 according to the Gregorian calendar, which had already been introduced in Italy and other Catholic countries; Isaac Newton was born on 25 December 1642 according to the Julian calendar still used in England and other Protestant countries. On the Gregorian calendar, the one we use today, Newton was born on 4 January 1643, while on the Julian calendar Galileo died right at the end of 1641. Either way, the two events did not take place in the same calendar year. But there is an equally noteworthy and genuine coincidence that results from taking Newton's birthday as 4 January 1643, in line with our modern calendar. In that case, he was born exactly 100 years after the publication of *De Revolutionibus,* which highlights how quickly science became established once it became part of the Renaissance.

Although the English Civil War disrupted many lives, as we have seen, over the next few years it largely passed by the quiet backwaters of Lincolnshire, and for three years Isaac Newton enjoyed the devoted attention of his widowed mother. But in 1645, just when he was old enough to appreciate this, she remarried and he was sent to live with his maternal grandparents. Almost literally ripped from his mother's arms and dumped in more austere surroundings at such a tender age, this scarred him mentally for life, although no unkindness was intended. Hannah was just being practical.

Like most marriages among the families of 'gentlemen' at the time (including Hannah's first marriage), this second marriage was a businesslike relationship rather than a love match. Hannah's new husband was a 63-year-old widower, Barnabas Smith, who needed a new partner and essentially chose Hannah from the available candidates (he was the Rector of North Witham, less than 3 kilometres from Woolsthorpe). The business side of the arrangement included the settling of a piece of land on young Isaac by the Rector, on condition that he lived away from the new matrimonial home. So while Hannah went off to North Witham, where she bore two daughters and a son before Barnabas Smith died in 1653, Isaac spent eight formative years as a solitary child in the care of elderly grandparents (they had married back in 1609 and must have been almost as old as Hannah's new husband), who seem to have been dutiful and strict, rather than particularly loving towards him.

The bad side of this is obvious enough, and clearly has a bearing on Isaac's development as a largely solitary individual who kept himself to himself and made few close friends. But the positive side is that he received an education.[1] Had his father lived, Isaac Newton would surely have followed in his footsteps

as a farmer; but to the Ayscough grandparents it was natural to send the boy to school (one suspects not least because it kept him out of the way). Although Isaac returned to his mother's house in 1653 when he was 10 and she was widowed for the second time, the seed had been sown, and when he was 12 he was sent to study at the grammar school in Grantham, about 8 kilometres from Woolsthorpe. While there, he lodged with the family of an apothecary, Mr Clark, whose wife had a brother, Humphrey Babington. Humphrey Babington was a Fellow of Trinity College in Cambridge, but spent most of his time at Boothby Pagnall, near Grantham, where he was rector.

Although Isaac seems to have been lonely at school, he was a good student and also showed an unusual ability as a model maker (echoing Hooke's skill), constructing such devices (much more than mere toys) as a working model of a windmill and flying a kite at night with a paper lantern attached to it, causing one of the earliest recorded UFO scares. In spite of his decent education (mostly the Classics, Latin and Greek), Newton's mother still expected him to take over the family farm when he became old enough, and in 1659 he was taken away from the school to learn (by practical experience) how to manage the land. This proved disastrous. More interested in books, which he took out to read in the fields, than in his livestock, Newton was fined several times for allowing his animals to damage other farmers' crops, and many stories of his absent-mindedness concerning his agricultural duties have come down to us, doubtless embroidered a little over the years. While Isaac was (perhaps to some extent deliberately) demonstrating his incompetence in this area, Hannah's brother William, the Cambridge graduate, was urging her to let the young man follow his natural inclinations and go up to the university. The combination of her brother's persuasion and the mess Isaac was making of her farm won her grudging

---

1 And in the longer term, he benefited in terms of financial security, inheriting (via his mother) not only his father's property, but also a share of the not inconsiderable property of Barnabas Smith and some of the property Hannah inherited from her parents. Isaac Newton never had to worry about money once he reached the age of 21 and came into the income from the land that had been settled

on him by Barnabas Smith as part of his wedding contract with Hannah.

acceptance of the situation, and in 1660 (the year of the Restoration) Isaac went back to school to prepare for admission to Cambridge. On the advice of (and, no doubt, partly thanks to the influence of) Humphrey Babington, he took up his place at Trinity College on 8 July 1661. He was then 18 years old-about the age people go to university today, but rather older than most of the young gentlemen entering Cambridge in the 1660s, when it was usual to go up to the university at the age if 14 or 15, accompanied by a servant.

Far from having his own servant, though, Isaac had to act as a servant himself. His mother would not tolerate more than the minimum expenditure on what she still regarded as a wasteful indulgence, and she allowed Isaac just £10 a year, although her own income at this time was in excess of £700 a year. Being a student's servant at this time (called a subsizar) could be extremely unpleasant and involve such duties as emptying the chamber pots of your master. It also had distinctly negative social overtones. But here Newton was lucky (or cunning); he was officially the student of Humphrey Babington, but Babington was seldom up at college and he was a friend who did not stress the master-servant relationship with Isaac. Even so, possibly because of his lowly status and certainly because of his introverted nature, Newton seems to have had a miserable time in Cambridge until early in, 1663, when he met and became friendly with Nicholas Wickins. They were both unhappy with their room-mates and decided to share rooms together, which they did in the friendliest way for the next 20 years. It is quite likely that Newton was a homosexual; the only close relationships he had were with men, although there is no evidence that these relationships were consummated physically (equally, there is no evidence that they weren't). This is of no significance to his scientific work, but may provide another clue to his secretive nature.

The scientific life began to take off once Newton decided pretty much to ignore the Cambridge curriculum, such as it was, and read what he wanted (including the works of Galileo and Descartes). In the 1660s, Cambridge was far from being a centre of academic excellence. Compared with Oxford it was a backwater and, unlike Oxford, had not benefited from any intimate contact with the Greshamites. Aristotle was still taught by rote, and the only thing a Cambridge education fitted anyone for was to be a competent priest or a bad doctor. But the first hint of what was to be came in 1663, when Henry Lucas endowed a professorship of mathematics in Cambridge-the first scientific professorship in the university (and the first new chair of any kind since 1540). The first holder of the Lucasian chair of mathematics was Isaac Barrow, previously a professor of Greek (which gives you some idea of where science stood in Cambridge at the time). The appointment was doubly significant-first because Barrow did teach some mathematics (his first course of lectures, in 1664, may well have been influential in stimulating Newton's interest in science) and then, as we have seen, because of what happened when he resigned from the position five years later.

According to Newton's own later account, it was during those five years, from 1663 to 1668, that he carried out most of the work for which he is now famous. I have already discussed his work on light and colour, which led to the famous row with Hooke. But there are two other key pieces of work which need to be put in context-Newton's invention of the mathematical techniques now known as calculus (which he called fluxions) and his work on gravity that led to the *Principia*.

Whatever the exact stimulus, by 1664 Newton was a keen (if unconventional) scholar and eager to extend his time at Cambridge. The way to do this was first to win one of the few scholarships available to an undergraduate and then to gain election, after a few years, to a Fellowship of the college. In April 1664, Newton achieved the essential first step, winning a scholarship in spite of his failure to follow the prescribed course of study, and almost certainly because of the influence of Humphrey Babington, by now a senior member of the college. The scholarship brought in a small income, provided for his keep and removed from him the stigma of being a subsizar; it also meant that after automatically receiving his BA in January 1665 (once you were up at Cambridge in those days it was impossible not to get a degree unless you chose, as many students did, to leave early), he could stay in residence, studying what he liked, until he became MA in 1668.

Newton was an obsessive character who threw himself body and soul into whatever project was at hand. He would forget to eat or sleep while studying or carrying out experiments, and carried out some truly alarming experiments on himself during his study of optics, gazing at the Sun for so long he nearly went blind and poking about in his eye with a bodkin (a fat, blunt, large-eyed needle) to study the coloured images resulting from this rough treatment. The same obsessiveness would surface in his later life, whether in his duties at the Royal Mint or in his many disputes with people such as Hooke and Gottfried Leibnitz, the other inventor of calculus. Although there is no doubt that Newton had the idea first, in the mid-1660s, there is also no doubt that Leibnitz (who lived from 1646 to 1716) hit on the idea independently a little later (Newton not having bothered to tell anyone of his work at the time) and that Leibnitz also came up with a more easily understood version. I don't intend to go into the mathematical details; the key thing about calculus is that it makes it possible to calculate accurately, from a known starting situation, things that vary as time passes, such as the position of a planet in its orbit. It would be tedious to go into all the details of the Newton-Leibnitz dispute; what matters is that they did develop calculus in the second half of the seventeenth century, providing scientists of the eighteenth and subsequent centuries with the mathematical tools they needed to study processes in which change occurs. Modern physical science simply would not exist without calculus.

Newton's great insights into these mathematical methods, and the beginnings of his investigation of gravity, occurred at a time when the routine of life in Cambridge was interrupted by the threat of plague. Not long after he graduated, the university was closed temporarily and its scholars dispersed to avoid the plague. In the summer of 1665, Newton returned to Lincolnshire, where he stayed until March *1666*. It then seemed safe to return to Cambridge, but with the return of warm weather the plague broke out once more, and in June he left again for the country, staying in Lincolnshire until April 1667, when the plague had run its course. While in Lincolnshire, Newton divided his time between Woolsthorpe and Babington's rectory in Boothby Pagnell, so there is no certainty about where the famous apple incident occurred (if indeed it really did occur at this time, as Newton claimed). But what is certain is that, in Newton's own words, written half a century later, 'in those days I was in the prime of my age for invention & minded Mathematicks & Philosophy more than at any time since'. At the end of 1666, in the midst of this inspired spell, Newton enjoyed his twenty-fourth birthday.

The way Newton later told the story, at some time during the plague years he saw an apple fall from a tree and wondered whether, if the influence of the Earth's gravity could extend to the top of the tree, it might extend all the way to the Moon. He then calculated that the force required to hold the Moon in its orbit and the force required to make the apple fall from the tree could both be explained by the Earth's gravity if the force fell off as one over the square of the distance from the centre of the Earth. The implication, carefully cultivated by Newton, is that he had the inverse square law by 1666, long before any discussions between Halley, Hooke and Wren. But Newton was a great one for rewriting history in his own favour, and the inverse square law emerged much more gradually than the story suggests. From the written evidence of Newton's own papers that can be dated, there is nothing about the Moon in the work on gravity he carried out in the plague years. What started him thinking about gravity was the old argument used by opponents of the idea that the Earth could be spinning on its axis, to the effect that if it were spinning it would break up and fly apart because of centrifugal force. Newton calculated the strength of this outward force at the surface of the Earth and compared it with the measured strength of gravity, showing that gravity at the surface of the Earth is hundreds of times stronger than the outward force, so the argument doesn't stand up. Then, in a document written some time after he returned to Cambridge (but certainly before 1670), he compared these forces to the 'endeavour of the Moon to recede from the centre of the Earth' and found that gravity *at the surface of the Earth* is about 4000 times stronger than the outward (centrifugal) force appropriate for the Moon moving in its orbit. This outward force would balance the force of the Earth's gravity if gravity fell off in accordance with an inverse square law, but Newton did not explicitly state this at the time. He also noted, though, from Kepler's laws, that the 'endeavours of receding from the Sun' of the planets in their orbits were inversely proportional to the squares of their distances from the Sun.

To have got this far by 1670 is still impressive, if not as impressive as the myth Newton later cultivated so assiduously. And remember that by this time, still not 30, he had essentially completed his work on light and on calculus. But the investigation of gravity now took a back seat as Newton turned to a new passion-alchemy. Over the next two decades, Newton devoted far more time and effort to alchemy than he had to all of the scientific work that we hold in such esteem today, but since this was a complete dead end there is no place for a detailed discussion of that work here.[2] He also had other distractions revolving around his

position at Trinity College and his own unorthodox religious convictions.

In 1667, Newton was elected to a minor Fellowship at Trinity, which would automatically become a major Fellowship when he became an MA in 1668. This gave him a further seven years to do whatever he liked, but involved making a commitment to orthodox religion-specifically, on taking up the Fellowship all new Fellows had to swear an oath that 'I will either set Theology as the object of my studies and will take holy orders when the time prescribed by these statutes arrives, or I will resign from the college.' The snag was that Newton was an Arian.[3] Unlike Bruno, he was not prepared to go to the stake for his beliefs, but nor was he prepared to compromise them by swearing on oath that he believed in the Holy Trinity, which he would be required to do on taking holy orders. To be an Arian in England in the late seventeenth century was not really a burning offence, but if it came out it would exclude Newton from public office and certainly from a college named after the Trinity. Here was yet another reason for him to be secretive and introverted; and in the early 1670s, perhaps searching for a loophole, Newton developed another of his long-term obsessions, carrying out detailed studies of theology (rivaling his studies of alchemy and helping to explain why he did no new science after he was 30). He was not saved by these efforts, though, but by a curious stipulation in the terms laid down by Henry Lucas for his eponymous chair.

Newton had succeeded Barrow as Lucasian professor in 1669, when he was 26. The curious stipulation, which ran counter to all the traditions of the university, was that any holder of the chair was barred from accepting a position in the Church requiring

---

2 For a popular account which focuses on this aspect of Newton's life, see *Isaac Newton: the last sorcerer*, by Michael White.

3 Newton followed the teaching of the fourth-century Alexandrian Arius. Arianism held that God is a unique being and that therefore Jesus was not truly divine. These were heretical ideas in the eyes of a Church based on the concept of the Holy Trinity.

residence outside Cambridge or 'the cure of souls'. In 1675, using this stipulation as an excuse, Newton got permission from Isaac Barrow (by now Master of Trinity) to petition the King for a dispensation for all Lucasian professors from the requirement to take holy orders. Charles II, patron of the Royal Society (where, remember, Newton was by now famous for his reflecting telescope and work on light) and enthusiast for science, granted the dispensation in perpetuity, 'to give all just encouragement to learned men who are & shall be elected to the said Professorship'. Newton was safe-on the King's dispensation, he would not have to take holy orders, and the college would waive the rule which said that he had to leave after his initial seven years as a Fellow were up.

In the middle of all the anxiety about his future in Cambridge Newton was also embroiled in the dispute with Hooke about priority over the theory of light, culminating in the 'shoulders of Giants' letter in 1675. We can now see why Newton regarded this whole business with such irritation-he was far more worried about his future position in Cambridge than about being polite to Hooke. Ironically, though, while Newton was distracted from following up his ideas about gravity, in 1674 Hooke had struck to the heart of the problem of orbital motion. In a treatise published that year, he discarded the idea of a balance of forces, some pushing inwards and some pushing outwards, to hold an object like the Moon in its orbit. He realized that the orbital motion results from the tendency of the Moon to move in a straight line, plus a *single* force pulling it towards the Earth. Newton, Huygens and everyone else still talked about a 'tendency to recede from the centre', or some such phrase, and the implication, even in Newton's work so far, was that something like Descartes's swirling vortices were responsible for pushing things back into their orbits in spite of this tendency to move outwards. Hooke also did away with the vortices, introducing the idea of what we would now call 'action at a distance'-gravity reaching out across *empty* space to tug on the Moon or the planets.

In 1679, after the dust from their initial confrontation had settled, Hooke wrote to Newton asking for his opinion on these ideas (which had already been published). It was Hooke who introduced Newton to the idea of action at a distance (which immediately appears, without comment, in all of Newton's subsequent work on gravity) and to the idea that an orbit is a straight line bent by gravity. But Newton was reluctant to get involved, and wrote to Hooke that:

I had for some years past been endeavouring to bend my self from Philosophy to other studies in so much y^r I have long grutched the time spent in y^t study ... I hope it will not be interpreted out of any unkindness to you or ye R. Society that I am backward in engaging myself in these matters.

In spite of this, Newton did suggest a way to test the rotation of the Earth. In the past, it had been proposed that the rotation of the Earth ought to show up if an object was dropped from a sufficiently tall tower, because the object would be left behind by the rotation and fall behind the tower. Newton pointed out that the top of the tower had to be moving faster than its base, because it was further from the centre of the Earth and had correspondingly more circumference to get round in 24 hours. So the dropped object ought to land in front of the tower. Rather carelessly, in a drawing to show what he meant, Newton carried through the trajectory of the falling body as if the Earth were not there, spiralling into the centre of the Earth under the influence of gravity. But he concluded the letter by saying:

But yet my affection to Philosophy being worn out, so that I am almost as little concerned about it as one tradesman uses to be about another man's trade or a country man about learning, I must acknowledge my self avers from spending that time in

writing about it. I think I can spend otherwise more to my own content.

But drawing that spiral drew Newton into more correspondence on 'Philosophy' whether he liked it or not. Hooke pointed out the error and suggested that the correct orbit followed by the falling object, assuming it could pass through the solid Earth without resistance, would be a kind of shrinking ellipse. Newton in turn corrected Hooke's surmise by showing that the object orbiting inside the Earth would not gradually descend to the centre along any kind of path, but would orbit indefinitely, following a path like an ellipse, but with the whole orbit shifting around as time passed. Hooke in turn replied to the effect that Newton's calculation was based on a force of attraction with 'an equal power at all Distances from the center ... But my supposition is that the Attraction always is in a duplicate proportion to the Distance from the Center Reciprocal', in other words, an inverse square law.

Newton never bothered to reply to this letter, but the evidence is that, in spite of his affection to philosophy being worn out, this was the trigger that stimulated him, in 1680, to *prove* (where Hooke and others could only surmise) that an inverse square law of gravity requires the planets to be in elliptical or circular orbits, and implied that comets should follow either elliptical or parabolic paths around the Sun.[4]

---

4 In fact, the hardest piece of mathematics in all of this is the proof that it is indeed correct to measure distances used in the inverse square law from the centre of the Earth or the Sun, with gravity acting as if all the mass were concentrated at a single point. Calculus makes this relatively straightforward, but Newton deliberately avoided using calculus in his published proofs, realizing that his peers would not accept the calculations unless they were couched in familiar language. Nobody knows if he did it all by calculus first and then translated into old-fashioned maths, but if he did, that is almost as impressive as if he did it the old-fashioned way to start with.

And that is why he was ready with the answer when Halley turned up on his doorstep in 1684.

It wasn't all plain sailing after that, but Halley's cajoling and encouragement following that encounter in Cambridge led first to the publication of a nine-page paper (in November 1684) spelling out the inverse square law work, and then to the publication in 1687 of Newton's epic three-volume work ***Philosophiae Naturalis Principia Mathematical*** in which he laid the foundations for the whole of physics, not only spelling out the implications of his inverse square law of gravity and the three laws of motion, which describe the behaviour of everything in the Universe, but making it clear that the laws of physics are indeed *universal* laws that affect everything. There was still time for one more glimpse of Newton's personality to surface-when Hooke complained that the manuscript (which he saw in his capacity at the Royal) gave him insufficient credit (a justifiable complaint, since he had achieved and passed on important insights even if he didn't have the skill to carry through the mathematical work as Newton could), Newton at first threatened to withdraw the third volume from publication and then went through the text before it was sent to the print-ers, savagely removing any reference at all to Hooke.

Apart from the mathematical brilliance of the way he fitted everything together, the reason why the ***Principia*** made such a big impact is because it achieved what scientists had been groping towards (without necessarily realizing it) ever since Copernicus-the real-ization that the world works on essentially mechanical principles that can be understood by human beings, and is not run in accordance with magic or the whims of capricious gods.

For Newton and many (but by no means all) of his contemporaries there was still a role for God as the architect of the whole thing, even a 'hands-on' architect who might interfere from time to time to ensure the smooth running of His creation. But it became increasingly clear to many who followed after Newton that however it started, once the Universe was up and running it ought to need no interference from outside. The analogy that is often used is with a clockwork mechanism. Think of a great church clock of Newton's time, not just with hands marking off the time, but with wooden figures that emerge from the interior on the hour, portraying a little tableau and striking the chimes with a hammer on a bell. A great complexity of surface activity, but all happening as a result of the tick-tock of a simple pendulum. Newton opened the eyes of scientists to the fact that the fundamentals of the Universe might be simple and understandable, in spite of its surface complexity.. He also had a clear grasp of the scientific method, and once wrote (to the French Jesuit Gaston Pardies):

> The best and safest method of philosophizing seems to be, first to inquire diligently into the properties of things, and to establish those properties by experiences and then to proceed more slowly to hypotheses for the explanation of them. For hypotheses should be employed only in explaining the properties of things, but not assumed in determining them; unless so far as they may furnish experiments.

In other words, science is about facts, not fancy.

The publication of the *Principia* marked the moment when science came of age as a mature intellectual discipline, putting aside most of the follies of its youth and settling down into grown-up investigation of the world. But it wasn't just because of Newton. He was a man of his times, putting clearly into words (and, crucially, into equations) ideas that were bubbling up all around, expressing more clearly than they could themselves what other scientists were already struggling to express. That is another reason why his book drew such a response-it struck a chord, because the time was ripe for such a summing up and laying of foundations. For almost every scientist who read the *Principia*, it must have been like the moment in C.

P. Snow's *The Search*, when T saw a medley of haphazard facts fall into line and order ... "but it's true," I said to myself. "It's very beautiful. And it's true.'"

Newton himself became a famous scientist, far beyond the circle of the Royal Society, as a result of the publication of the *Principia*, The philosopher John Locke, a friend of Newton, wrote of the book:

> The incomparable Mr Newton has shewn, how far mathematicks, applied to some Parts of Nature, may upon Principles that Matter of Fact justifies, carry us in the knowledge of some, as I may so call them, particular Provinces of the Incomprehensible Universe.

But in 1687, Newton had stopped being a scientist (he would be 45 at the end of that year and had long since lost his affection for philosophy). True, his *Opticks* would be published at the beginning of the eighteenth century-but that was old work, sat upon until Hooke died and it could be published without him getting a chance to comment on it or claim any credit for his own work on light. But the status that resulted from the *Principia* may have been one of the factors that encouraged Newton to become a public figure in another sense, and although the rest of his life story has little direct bearing on science, it is worth sketching out just how much he achieved apart from his scientific work.

Newton's first move into the political limelight came early in 1687, alter the *Principia* was off his hands and being seen through printing by Halley. James II had succeeded his brother in 1685 and, after a cautious start to his reign, by 1687 he was starting to throw his weight around. Among other things, he tried to extend Catholic influence over the University of Cambridge. Newton, by now a senior Fellow at Trinity (and perhaps influenced by fears of what might happen to him as an Arian under a Catholic regime), was one of the leaders of the opposition to these moves in Cambridge and was one of nine Fellows who had to appear before the notorious Judge Jeffreys to defend their stand. When James was removed from the throne at the end of 1688 and replaced early in 1689 by William of Orange (a grandson of Charles I) and his wife Mary (a daughter of James II) in the so-called Glorious Revolution,[5] Newton became one of two Members of Parliament sent to London by the University. Although far from active in Parliament, and not offering himself for re-election when the Parliament (having done its job of legalizing the takeover by William and Mary) was dissolved early in 1690, the taste of London life and participation in great events increased Newton's growing dissatisfaction with Cambridge. Although he threw himself into his alchemical work in the early 1690s, in 1693 he seems to have suffered a major nervous breakdown brought on by years of overwork, the strain of concealing his unorthodox religious views and (possibly) the breakup of a close friendship he had had over the previous three years with a young mathematician from Switzerland, Nicholas Fatio de Duillier (usually known as Fatio). When Newton recovered, he sought almost desperately for some way to leave Cambridge, and when in 1696 he was offered the Wardenship of the Royal Mint (by Charles Montague, a former Cambridge student, born in 1661, who knew Newton

and was by now Chancellor of the Exchequer, but also found time to serve as President of the Royal Society from 1695 to 1698), he leapt at the chance.

The Wardenship was actually the number two job at the Mint, and could be treated as a sinecure. But since the then Master of the Mint effectively treated his own post as a sinecure, Newton had a chance to get his hands on the levers of power. In his obsessive way he took the place over, seeing through a major recoinage and cracking down on counterfeiters with ferocity and cold-blooded ruthlessness (the punishment was usually hanging, and Newton became a magistrate in order to make sure that the law was on his side). When the Master died in 1699, Newton took over the post-the only time in the long history of the Mint that the Warden moved up in this way. Newton's great success at the Mint encouraged him (probably at the urging of Montague, who was by now Lord Halifax and later became the Earl of Halifax) to stand again for Parliament, which he did successfully in 1701, serving until May 1702, when William II died (Mary had pre-deceased him in 1694) and Parliament was dissolved. William was succeeded by Anne, the second daughter of James II, who, before and during her twelve-year reign, was greatly influenced by Halifax. During the election campaign of 1705, she knighted both Newton, Halifax's protégé, and Halifax's brother, in the hope that the honour would encourage voters to support them.

It did them no good-Halifax's party as a whole lost the election, as did Newton as an individual, and Newton, now in his sixties, never stood again. But the story is worth telling, since many people think that Newton received his knighthood for science and some think that it was a reward for his work at the Mint, but the truth is that it was a rather grubby bit of political opportunism by Halifax as part of his attempt to win the election of 1705.

---

5 William and Mary actually gained the throne as a result of a full-scale invasion of Britain, seizing London by force, even if it was largely bloodless and welcomed by many. But history is written by the victors, and Glorious Revolution sounds so much better than Invasion if you want to keep the populace happy. The most significant feature of the 'revolution' (almost justifying the name) was that it tilted the balance of political power in Britain away from the King and in favour of Parliament, without whom William and Mary could not have succeeded.

Section III
Philosophy of Science

Introduction

By John Oakes

The philosopher asks how we know what we know. The branch of philosophy that studies how we acquire reliable knowledge is **epistemology**. One way of acquiring knowledge is by **induction**. To induce something is to discover a general truth by observation of specific examples. By observation, we can discover/induce that glowing red/orange metal objects are hot. We can induce that the sun rises every day (even though we know from other evidence that it is actually the spinning of the earth that reveals the relatively immobile sun to us every day!). Induction involves looking into the past to discover a pattern. By definition, one cannot induce the future.

Another way of gaining knowledge is by **deduction**. Deduction goes from the general to the specific—from law to application of that law. Deduction involves logical inference. For example, by observation I might induce that all swans are white. Let us assume, for now, that all swans are white. Using that premise, I can deduce that a swan I have not yet observed will be white. Deduction can be used to make predictions. Scientific laws are the result of many observations and are the result of induction. Scientific theories are explanations and can be used deductively to predict behaviours not yet observed. Scientific thinking can be viewed as a balanced use of induction and deduction to acquire knowledge of the world.

Philosophers of science can be broadly grouped into rationalists and empiricists. The **rationalist** says that we know things are true because our mind tells us so. The rationalist believes that we cannot trust our senses to reveal truth. Instead, we can use logic and reasoning to determine what is true. The rationalist believes the world can be fully described using a fairly small number of mathematical axioms, from which we can derive all the laws of nature. The Greek natural philosophers had a rationalist approach to the world. Thomas Aquinas (1224–1274) was an important early Western rationalist. He sought to apply the logic of Aristotle to explain Christian theology. You will read a selection from *Discourse on Method*, by René Descartes. Famously, Descartes said, "I think, therefore I am." His philosophy of science can be described as rationalism. To Descartes, the ideal scientist is the theoretician. He said "The need to experiment is an expression of the failure of the ideal." To rationalists like Descartes, the theoretician is the ideal scientist. To them, mathematics—not observation—is the most important tool of the scientist.

Most of the early philosophers of science were empiricists. The **empiricist** says that we learn through our senses. Experience tells us what is true. The empiricist relies on observation to discover knowledge. He or she does not trust rational thought to produce reliable knowledge. The first important philosopher of science was Roger Bacon (1214–1292). His approach to knowledge was empirical. In his *Opus Majus*, he proposed an empirical approach to natural philosophy, with experimentation as the heart of this method. To acquire knowledge of the physical world, one must use "external experience, aided by instruments, made precise by mathematics." His was the first statement that the physical world can be well described by mathematics, but his approach was empirical. His commitment to empiricism is revealed by another statement in *Opus Majus*. "He who wishes to rejoice without doubt in regard to the truths underlying phenomena must know how to devote himself to experiment. For authors write many statements, and people believe them through [deductive] reasoning, which they formulate without experience. Their reasoning is wholly false."

One of the most important early empiricists was William of Ockham (1285–1349). He said that "nothing is assumed as evident unless it is known *per se*, evident by experience or proved by authority of scripture." Here he saw the place for deduction (known *per se*) and induction in our means of discovering knowledge.

The most influential philosopher of science during the scientific revolution was Francis Bacon (1561–1626, no relation to Roger Bacon). His was a practical empiricism. With a bit of exaggeration, some have described Francis Bacon as the inventor of the scientific method. His most important work in the philosophy of science was *Novum Organum*. In it he said "The end of our foundation is the knowledge of causes, and the secret motions of things, and the enlarging of the bounds of human empire, to the affecting of all things possible." He proposed to acquire this knowledge through experiment, not through rational thought. He saw the potential for great practical use of science. "Knowledge ought to bear fruit in work, that science ought to be applicable to industry, that man ought to organize themselves as a sacred duty to improve and transform the conditions of life."

As you read selections on the philosophy of science, you should ask yourself how the author believes knowledge of the physical world is gained. You will see Galileo defending not just his heliocentric view of the universe, but the scientific process itself. Galileo defends an empirical approach to knowledge of the world. As you read the piece by Cushing, ask yourself where David Hume, John Mill, and Karl Popper fit into the rationalist/empiricist continuum. You will be asked to consider the problem of induction. Can we use induction to make predictions? Is verification by experiment possible? What is the demarcation between science and nonscience? Last, you will read a piece by Thomas Kuhn. This influential historian of science coined the word "paradigm." He viewed the history of scientific progress, not as a gradual increase in knowledge, but as a series of "revolutions." Scientific "truth" is determined more by the accepted world view than by any kind of pure scientific method. As you consider what you have already learned about science, ask yourself to what extent you find yourself agreeing with Thomas Kuhn.

Letter to the Duchess Christina

By Galileo Galilei

Editor's Note: The following is an edited version of Galileo's original letter, with portions removed for brevity.

The reason produced for condemning the opinion that the earth moves and the sun stands still in many places in the Bible. One may read that the sun moves and the earth stands still. Since the Bible cannot err, it follows as a necessary consequence that anyone takes a erroneous and heretical position who maintains that the sun is inherently motionless and the earth movable.

With regard to this argument, I think in the first place that it is very pious to say and prudent to affirm that the holy Bible can never speak untruth—whenever its true meaning is understood. But I believe nobody will deny that it is often very abstruse, and may say things which are quite different from what its bare words signify. Hence in expounding the Bible if one were always to confine oneself to the unadorned grammatical meaning, one might fall into error. Not only contradictions and propositions far from true might thus be made to appear in the Bible, but even grave heresies and follies. Thus it would be necessary to assign to God feet, hands and eyes—as well as corporeal and human affections—such as anger, repentance, hatred, and sometimes even the forgetting of things past and ignorance of those to come. These propositions uttered by the Holy Ghost were set down in that manner by the sacred scribes in order to accommodate them to the capacities, Of the common people, who are rude and unlearned. For the sake of those who deserve to be separated from the herd, it is necessary that wise expositors should produce the true senses of such passages, together with the special reasons for which they were set down in these words. This doctrine is so widespread and so definite with all theologians that it would be superfluous to adduce evidence for it.

Hence I think that I may reasonably conclude that whenever the Bible has occasion to speak of any physical conclusion (especially those which are very abstruse and hard to understand), the rule has been observed of avoiding confusion in the minds of the common people which would render them contumacious toward the higher mysteries. Now the Bible, merely to condescend to popular capacity, has not hesitated to obscure some very important pronouncements, attributing to God himself some qualities extremely remote from (and even contrary to) His essence. Who, then, would positively declare that this principle has been set aside, and the Bible has confined itself rigorously to the bare and restricted sense of its words, when speaking but casually of the earth, of water, of the sun, or of any other created thing? Especially in view of the fact that these things in no way concern the primary purpose of the sacred writings, which is the service of God and the salvation of souls—matters infinitely beyond the comprehension of the common people.

This being granted, I think that in discussions of physical problems we ought to begin not from the authority of scriptural passages but from sense experiences and necessary demonstrations. For the holy Bible and the phenomena of nature proceed alike from the divine Word the former as the dictate of the Holy Ghost and the latter as the observant executrix of God's commands. It is necessary for the Bible, in order to be accommodated to the understanding of every man, to speak many things which appear to differ from the absolute truth so far as the bare meaning of the words is concerned. But Nature, on the other hand, is inexorable and immutable; she never transgresses the laws imposed upon her, or cares a whit whether her abstruse reasons and methods of operation are understandable to men. For that reason it appears that nothing physical which sense experience sets before our eyes, or which necessary demonstrations prove to us, ought to be called in question (much less condemned) upon the testimony of biblical passages which may have some different meaning beneath their words. For the Bible is not chained in every expression to conditions as strict as those which govern all physical effects; nor is God any less excellently revealed in Nature's actions than in the sacred statements of the Bible. Perhaps this is what Tertullian meant by these words:

"We conclude that God is known first through Nature, and then again, more particularly, by doctrine, by Nature in His works, and by doctrine in His revealed word."

From this I do not mean to infer that we need not have an extraordinary esteem for the passages of holy Scripture. On the contrary, having arrived at any certainties in physics, we ought to utilize these as the most appropriate aids in the true exposition of the Bible and in the investigation of those meanings which are necessarily contained therein, for these must be concordant with demonstrated truths. I should judge that the authority of the Bible was designed to persuade men of those articles and propositions which, surpassing all human reasoning could not be made credible by science, or by any other means than through the very mouth of the Holy Spirit.

Yet even in those propositions which are not matters of faith, this authority ought to be preferred over that of all human writings which are supported only by bare assertions or probable arguments, and not set forth in a demonstrative way. This I hold to be necessary and proper to the same extent that divine wisdom surpasses all human judgment and conjecture.

But I do not feel obliged to believe that the same God who has endowed us with senses, reason and intellect has intended us to forego their use and by some other means to give us knowledge which we can attain by them. He would not require us to deny sense and reason in physical matters which are set before our eyes and minds by direct experience or necessary demonstrations. This must be especially true in those sciences of which but the faintest trace (and that consisting of conclusions) is to be found in the Bible. Of astronomy; for instance, so little is found that none of the planets except Venus are so much as mentioned, and this only once or twice under the name of "Lucifer." If the sacred scribes had had any intention of teaching people certain arrangements and motions of the heavenly bodies, or had they wished us to derive such knowledge from the Bible, then in my opinion they would not have spoken of these matters so sparingly in comparison with the infinite number of admirable conclusions which are demonstrated in that science. Far from pretending to teach us the constitution and motions of the heavens and other stars, with their shapes, magnitudes, and distances, the authors of the Bible intentionally forbore to speak of these things, though all were quite well known to them. Such is the opinion of the holiest and most learned Fathers...

From these things it follows as a necessary consequence that, since the Holy Ghost did not intend to teach us whether heaven moves or stands still, whether its shape is spherical or like a discus or extended in a plane, nor whether the earth is located at its center or off to one side, then so much the less was it intended to settle for us any other conclusion of the same kind. And the motion or rest of the earth and the sun is so closely linked with the things just named, that without a determination of the one, neither side can be taken in the other matters. Now if the Holy Spirit has purposely neglected to teach us propositions of this sort as irrelevant to the highest goal (that is, to our salvation), how can anyone affirm that it is obligatory to take sides on them, that one belief is required by faith, while the other side is erroneous? Can an opinion be heretical and yet have no concern with the salvation of souls? Can the Holy Ghost be asserted not to have intended teaching us something that does concern our salvation? I would say here something that was heard from an ecclesiastic of the most eminent degree: "That the intention of the Holy Ghost is to teach us how one goes to heaven, not how heaven goes."

If in order to banish the opinion in question from the world it were sufficient to stop the mouth of a single man-as perhaps those men persuade themselves who, measuring the minds of others by their own, think it impossible that this doctrine should be able to continue to find adherents-then that would be very easily done. But things stand otherwise. To carry out such a decision it would be necessary not only to prohibit the book of Copernicus and the writings of other authors who follow the same opinion, but to ban the whole science of astronomy. Furthermore, it would be necessary to forbid men to look at the heavens, in order that they might not see Mars and Venus sometimes quite near the earth and sometimes very distant, the variation being so great that Venus is forty times and Mars sixty times as large at one time as at another. And it would be necessary to prevent Venus being seen round at one time and forked at another, with very thin horns; as well as many other sensory observations which can never be reconciled with the Ptolemaic system in any way, but are very strong arguments for the Copernican. And to ban Copernicus now that his doctrine is daily reinforced by many new observations and by the learned applying themselves to the reading of his book, after this opinion has been allowed and tolerated for these many years during which it was less followed and less confirmed, would seem in my judgment to be a contravention of truth, and an attempt to hide and suppress her the more as she revealed herself the more clearly and plainly. Not to abolish and censure his whole book, but only to condemn as erroneous this particular proposition, would (if I am not mistaken) be a still greater detriment to the minds of men, since it would afford them occasion to see a proposition proved that it was heresy to believe. And to prohibit the whole science would be to censure a hundred passages of holy Scripture which teach us that the glory and greatness of Almighty God are marvelously discerned in all his works and divinely read in the open book of heaven. For let no one believe that reading the lofty concepts written in that book leads to nothing further than the mere seeing of the splendor of the sun and the stars and their rising and setting, which is as far as the eyes of brutes and of the vulgar can penetrate. Within its pages are couched mysteries so profound and concepts so sublime that the vigils, labors, and studies of hundreds upon hundreds of the most acute minds have still not pierced them, even after the continual investigations for thousands of years.

Discourse on Method

By Rene Descartes

PREFATORY NOTE BY THE AUTHOR

If this Discourse appear too long to be read at once, it may be divided into six Parts: and, in the first, will be found various considerations touching the Sciences; in the second, the principal rules of the Method which the Author has discovered, in the third, certain of the rules of Morals which he has deduced from this Method; in the fourth, the reasonings by which he establishes the existence of God and of the Human Soul, which are the foundations of his Metaphysic; in the fifth, the order of the Physical questions which he has investigated, and, in particular, the explication of the motion of the heart and of some other difficulties pertaining to Medicine, as also the difference between the soul of man and that of the brutes; and, in the last, what the Author believes to be required in order to greater advancement in the investigation of Nature than has yet been made, with the reasons that have induced him to write.

PART ONE

Good sense is, of all things among men, the most equally distributed; for every one thinks himself so abundantly provided with it, that those even who are the most difficult to satisfy in everything else, do not usually desire a larger measure of this quality than they already possess. And in this it is not likely that all are mistaken the conviction is rather to be held as testifying that the power of judging aright and of distinguishing truth from error, which is properly what is called good sense or reason, is by nature equal in all men; and that the diversity of our opinions, consequently, does not arise from some being endowed with a larger share of reason than others, but solely from this, that we conduct our thoughts along different ways, and do not fix our attention on the same objects. For to be possessed of a vigorous mind is not enough; the prime requisite is rightly to apply it. The greatest minds, as they are capable of the highest excellences, are open likewise to the greatest aberrations; and those who travel very slowly may yet make far greater progress, provided they keep always to the straight road, than those who, while they run, forsake it.

For myself, I have never fancied my mind to be in any respect more perfect than those of the generality; on the contrary, I have often wished that I were equal to some others in promptitude of thought, or in clearness and distinctness of imagination, or in fullness and readiness of memory. And besides these, I know of no other qualities that contribute to the perfection of the mind; for as to the reason or sense, inasmuch as it is that alone which constitutes us men, and distinguishes us from the brutes, I am disposed to believe that it is to be found complete in each individual; and on this point to adopt the common opinion of philosophers, who say that the difference of greater and less holds only among the accidents, and not among the forms or natures of individuals of the same species.

I will not hesitate, however, to avow my belief that it has been my singular good fortune to have very early in life fallen in with certain tracks which have conducted me to considerations and maxims, of which I have formed a method that gives me the means, as I think, of gradually augmenting my knowledge, and of raising it by little and little to the highest point which the mediocrity of my talents and the brief duration of my life will permit me to reach. For I have already reaped from it such fruits that, although I have been accustomed to think lowly enough of myself, and although when I look with the eye of a philosopher at the varied courses and pursuits of mankind at large, I find scarcely one which does not appear in vain and useless, I nevertheless derive the highest satisfaction from the progress I conceive myself to have already made in the search after truth, and cannot help entertaining such expectations of the future as to believe that if, among the occupations of men as men, there is any one really excellent and important, it is that which I have chosen.

After all, it is possible I may be mistaken; and it is but a little copper and glass, perhaps, that I take for gold and diamonds. I know how very liable we are to delusion in what relates to ourselves, and also how much the judgments of our friends are to be suspected when given in our favor. But I shall endeavor in this discourse to describe the paths I have followed, and to delineate my life as in a picture, in order that each one may also be able to judge of them for himself, and that in the general opinion entertained of them, as gathered from current report, I myself may have a new

help towards instruction to be added to those I have been in the habit of employing.

My present design, then, is not to teach the method which each ought to follow for the right conduct of his reason, but solely to describe the way in which I have endeavored to conduct my own. They who set themselves to give precepts must of course regard themselves as possessed of greater skill than those to whom they prescribe; and if they err in the slightest particular, they subject themselves to censure. But as this tract is put forth merely as a history, or, if you will, as a tale, in which, amid some examples worthy of imitation, there will be found, perhaps, as many more which it were advisable not to follow, I hope it will prove useful to some without being hurtful to any, and that my openness will find some favor with all.

From my childhood, I have been familiar with letters; and as I was given to believe that by their help a clear and certain knowledge of all that is useful in life might be acquired, I was ardently desirous of instruction. But as soon as I had finished the entire course of study, at the close of which it is customary to be admitted into the order of the learned, I completely changed my opinion. For I found myself involved in so many doubts and errors, that I was convinced I had advanced no farther in all my attempts at learning, than the discovery at every turn of my own ignorance. And yet I was studying in one of the most celebrated schools in Europe, in which I thought there must be learned men, if such were anywhere to be found. I had been taught all that others learned there; and not contented with the sciences actually taught us, I had, in addition, read all the books that had fallen into my hands, treating of such branches as are esteemed the most curious and rare. I knew the judgment which others had formed of me; and I did not find that I was considered inferior to my fellows, although there were among them some who were already marked out to fill the places of our instructors. And, in fine, our age appeared to me as flourishing, and as fertile in powerful minds as any preceding one. I was thus led to take the liberty of judging of all other men by myself, and of concluding that there was no science in existence that was of such a nature as I had previously been given to believe.

I still continued, however, to hold in esteem the studies of the schools. I was aware that the languages taught in them are necessary to the understanding of the writings of the ancients; that the grace of fable stirs the mind; that the memorable deeds of history elevate it; and, if read with discretion, aid in forming the judgment; that the perusal of all excellent books is, as it were, to interview with the noblest men of past ages, who have written them, and even a studied interview, in which are discovered to us only their choicest thoughts; that eloquence has incomparable force and beauty; that poesy has its ravishing graces and delights; that in the mathematics there are many refined discoveries eminently suited to gratify the inquisitive, as well as further all the arts an lessen the labour of man; that numerous highly useful precepts and exhortations to virtue are contained in treatises on morals; that theology points out the path to heaven; that philosophy affords the means of discoursing with an appearance of truth on all matters, and commands the admiration of the more simple; that jurisprudence, medicine, and the other sciences, secure for their cultivators honors and riches; and, in fine, that it is useful to bestow some attention upon all, even upon those abounding the most in superstition and error, that we may be in a position to determine their real value, and guard against being deceived.

But I believed that I had already given sufficient time to languages, and likewise to the reading of the writings of the ancients, to their histories and fables. For to hold converse with those of other ages and to travel, are almost the same thing. It is useful to know something of the manners of different nations, that we may be enabled to form a more correct judgment regarding our own, and be prevented from thinking that everything contrary to our customs is ridiculous and irrational, a conclusion usually come to by those whose experience has been limited to their own country. On the other hand, when too much time is occupied in traveling, we become strangers to our native country; and the over curious in the customs of the past are generally ignorant of those of the present. Besides, fictitious narratives lead us to imagine the possibility of many events that are impossible; and even the most faithful histories, if they do not wholly misrepresent matters, or exaggerate their importance to render the account of them more worthy of perusal, omit, at least, almost always the meanest and least striking of the attendant circumstances; hence it happens that the remainder does not represent the truth, and that such as regulate their conduct by examples drawn from this source, are apt to fall into the extravagances of the knight-errants of romance, and to entertain projects that exceed their powers.

I esteemed eloquence highly, and was in raptures with poesy; but I thought that both were gifts of nature rather than fruits of study. Those in whom the faculty of reason is predominant, and who most skillfully dispose their thoughts with a view to render them clear and intelligible, are always the best able to persuade others of the truth of what they lay down, though they should speak only in the language of Lower Brittany, and be wholly ignorant of the rules of rhetoric; and those whose minds are stored with the most agreeable fancies, and who can give expression to them with the greatest embellishment and harmony, are still the best poets, though unacquainted with the art of poetry.

I was especially delighted with the mathematics, on account of the certitude and evidence of their reasonings; but I had not as yet a precise knowledge of their true use; and thinking that they but contributed to the advancement of the mechanical arts, I was astonished that foundations, so strong and solid, should have had no loftier superstructure reared on them. On the other hand, I compared the disquisitions of the ancient moralists to very towering and magnificent palaces with no better foundation than sand and mud: they laud the virtues very highly, and exhibit them as estimable far above anything on earth; but they give us no adequate criterion of virtue, and frequently that which they designate with so fine a name is but apathy, or pride, or despair, or parricide.

I revered our theology, and aspired as much as any one to reach heaven: but being given assuredly to understand that the way is not less open to the most ignorant than to the most learned, and that the revealed truths which lead to heaven are above our comprehension, I did not presume to subject them to the impotency of my reason; and I thought that in order competently to undertake their examination, there was need of some special help from heaven, and of being more than man.

Of philosophy I will say nothing, except that when I saw that it had been cultivated for many ages by the most distinguished men, and that yet there is not a single matter within its sphere which is not still in dispute, and nothing, therefore, which is above doubt, I did not presume to anticipate that my success would be greater in it than that of others; and further, when I considered the number of conflicting opinions touching a single matter that may be upheld by learned men, while there can be but one true, I reckoned as well-nigh false all that was only probable.

As to the other sciences, inasmuch as these borrow their principles from philosophy, I judged that no solid superstructures could be reared on foundations so infirm; and neither the honor nor the gain held out by them was sufficient to determine me to their cultivation: for I was not, thank Heaven, in a condition which compelled me to make merchandise of science for the bettering of my fortune; and though I might not profess to scorn glory as a cynic, I yet made very slight account of that honor which I hoped to acquire only through fictitious titles. And, in fine, of false sciences I thought I knew the worth sufficiently to escape being deceived by the professions of an alchemist, the predictions of an astrologer, the impostures

of a magician, or by the artifices and boasting of any of those who profess to know things of which they are ignorant.

For these reasons, as soon as my age permitted me to pass from under the control of my instructors, I entirely abandoned the study of letters, and resolved no longer to seek any other science than the knowledge of myself, or of the great book of the world. I spent the remainder of my youth in traveling, in visiting courts and armies, in holding intercourse with men of different dispositions and ranks, in collecting varied experience, in proving myself in the different situations into which fortune threw me, and, above all, in making such reflection on the matter of my experience as to secure my improvement. For it occurred to me that I should find much more truth in the reasonings of each individual with reference to the affairs in which he is personally interested, and the issue of which must presently punish him if he has judged amiss, than in those conducted by a man of letters in his study, regarding speculative matters that are of no practical moment, and followed by no consequences to himself, farther, perhaps, than that they foster his vanity the better the more remote they are from common sense; requiring, as they must in this case, the exercise of greater ingenuity and art to render them probable. In addition, I had always a most earnest desire to know how to distinguish the true from the false, in order that I might be able clearly to discriminate the right path in life, and proceed in it with confidence.

It is true that, while busied only in considering the manners of other men, I found here, too, scarce any ground for settled conviction, and remarked hardly less contradiction among them than in the opinions of the philosophers. So that the greatest advantage I derived from the study consisted in this, that, observing many things which, however extravagant and ridiculous to our apprehension, are yet by common consent received and approved by other great nations, I learned to entertain too decided a belief in regard to nothing of the truth of which I had been persuaded merely by example and custom; and thus I gradually extricated myself from many errors powerful enough to darken our natural intelligence, and incapacitate us in great measure from listening to reason. But after I had been occupied several years in thus studying the book of the world, and in essaying to gather some experience, I at length resolved to make myself an object of study, and to employ all the powers of my mind in choosing the paths I ought to follow, an undertaking which was accompanied with greater success than it would have been had I never quitted my country or my books.

PART TWO

I was then in Germany, attracted thither by the wars in that country, which have not yet been brought to a termination; and as I was returning to the army from the coronation of the emperor, the setting in of winter arrested me in a locality where, as I found no society to interest me, and was besides fortunately undisturbed by any cares or passions, I remained the whole day in seclusion, with full opportunity to occupy my attention with my own thoughts. Of these one of the very first that occurred to me was, that there is seldom so much perfection in works composed of many separate parts, upon which different hands had been employed, as in those completed by a single master. Thus it is observable that the buildings which a single architect has planned and executed, are generally more elegant and commodious than those which several have attempted to improve, by making old walls serve for purposes for which they were not originally built. Thus also, those ancient cities which, from being at first only villages, have become, in course of time, large towns, are usually but ill laid out compared with the regularity constructed towns which a professional architect has freely planned on an open plain; so that although the several buildings of the former may

often equal or surpass in beauty those of the latter, yet when one observes their indiscriminate juxtaposition, there a large one and here a small, and the consequent crookedness and irregularity of the streets, one is disposed to allege that chance rather than any human will guided by reason must have led to such an arrangement. And if we consider that nevertheless there have been at all times certain officers whose duty it was to see that private buildings contributed to public ornament, the difficulty of reaching high perfection with but the materials of others to operate on, will be readily acknowledged. In the same way I fancied that those nations which, starting from a semi-barbarous state and advancing to civilization by slow degrees, have had their laws successively determined, and, as it were, forced upon them simply by experience of the hurtfulness of particular crimes and disputes, would by this process come to be possessed of less perfect institutions than those which, from the commencement of their association as communities, have followed the appointments of some wise legislator. It is thus quite certain that the constitution of the true religion, the ordinances of which are derived from God, must be incomparably superior to that of every other. And, to speak of human affairs, I believe that the preeminence of Sparta was due not to the goodness of each of its laws in particular, for many of these were very strange, and even opposed to good morals, but to the circumstance that, originated by a single individual, they all tended to a single end. In the same way I thought that the sciences contained in books (such of them at least as are made up of probable reasonings, without demonstrations), composed as they are of the opinions of many different individuals massed together, are farther removed from truth than the simple inferences which a man of good sense using his natural and unprejudiced judgment draws respecting the matters of his experience. And because we have all to pass through a state of infancy to manhood, and have been of necessity, for a length of time, governed by our desires and preceptors (whose dictates were frequently conflicting, while neither perhaps always counseled us for the best), I farther concluded that it is almost impossible that our judgments can be so correct or solid as they would have been, had our reason been mature from the moment of our birth, and had we always been guided by it alone.

It is true, however, that it is not customary to pull down all the houses of a town with the single design of rebuilding them differently, and thereby rendering the streets more handsome; but it often happens that a private individual takes down his own with the view of erecting it anew, and that people are even sometimes constrained to this when their houses are in danger of falling from age, or when the foundations are insecure. With this before me by way of example, I was persuaded that it would indeed be preposterous for a private individual to think of reforming a state by fundamentally changing it throughout, and overturning it in order to set it up amended; and the same I thought was true of any similar project for reforming the body of the sciences, or the order of teaching them established in the schools: but as for the opinions which up to that time I had embraced, I thought that I could not do better than resolve at once to sweep them wholly away, that I might afterwards be in a position to admit either others more correct, or even perhaps the same when they had undergone the scrutiny of reason. I firmly believed that in this way I should much better succeed in the conduct of my life, than if I built only upon old foundations, and leaned upon principles which, in my youth, I had taken upon trust. For although I recognized various difficulties in this undertaking, these were not, however, without remedy, nor once to be compared with such as attend the slightest reformation in public affairs. Large bodies, if once overthrown, are with great difficulty set up again, or even kept erect when once seriously shaken, and the fall of such is always disastrous. Then if there are any imperfections in the constitutions of states (and that many such exist the diversity of constitutions is alone sufficient to assure us), custom has without doubt materially smoothed their inconveniences, and has even managed to steer altogether clear of, or insensibly corrected a number which sagacity could not have provided against with equal effect; and, in fine, the defects are almost always more tolerable than the change necessary for their removal; in the same manner that highways which wind among mountains, by being much frequented, become gradually so smooth and commodious, that it is much better to follow them than to seek a straighter path by climbing over the tops of rocks and descending to the bottoms of precipices.

Hence it is that I cannot in any degree approve of those restless and busy meddlers who, called neither by birth nor fortune to take part in the management of public affairs, are yet always projecting reforms; and if I thought that this tract contained aught which might justify the suspicion that I was a victim of such folly, I would by no means permit its publication. I have never contemplated anything higher than the reformation of my own opinions, and basing them on a foundation wholly my own. And although my own satisfaction with my work has led me to present here a draft of it, I do not by any means therefore recommend to every one else to make a similar attempt. Those whom God has endowed with a larger measure of genius will entertain, perhaps, designs still more exalted; but for the many I am much afraid lest even the present undertaking be more than they can safely venture to imitate. The single design to strip one's self of all past beliefs is one that ought not to be taken by every one. The majority of men is composed of two classes, for neither of which would this be at all a befitting resolution: in the first place, of those who with more than a due confidence in their own powers, are precipitate in their judgments and want the patience requisite for orderly and circumspect thinking; whence it happens, that if men of this class once take the liberty to doubt of their accustomed opinions, and quit the beaten highway, they will never be able to thread the byway that would lead them by a shorter course, and will lose themselves and continue to wander for life; in the second place, of those who, possessed of sufficient sense or modesty to determine that there are others who excel them in the power of discriminating between truth and error, and by whom they may be instructed, ought rather to content themselves with the opinions of such than trust for more correct to their own reason.

For my own part, I should doubtless have belonged to the latter class, had I received instruction from but one master, or had I never known the diversities of opinion that from time immemorial have prevailed among men of the greatest learning. But I had become aware, even so early as during my college life, that no opinion, however absurd and incredible, can be imagined, which has not been maintained by some on of the philosophers; and afterwards in the course of my travels I remarked that all those whose opinions are decidedly repugnant to ours are not in that account barbarians and savages, but on the contrary that many of these nations make an equally good, if not better, use of their reason than we do. I took into account also the very different character which a person brought up from infancy in France or Germany exhibits, from that which, with the same mind originally, this individual would have possessed had he lived always among the Chinese or with savages, and the circumstance that in dress itself the fashion which pleased us ten years ago, and which may again, perhaps, be received into favor before ten years have gone, appears to us at this moment extravagant and ridiculous. I was thus led to infer that the ground of our opinions is far more custom and example than any certain knowledge. And, finally, although such be the ground of our opinions, I remarked that a plurality of suffrages is no guarantee of truth where it is at all of difficult discovery, as in such cases it is much more likely that it will be found by one than by many. I could, however, select from the crowd no one whose opinions seemed worthy of preference, and thus I found myself constrained, as it were, to use my own reason in the conduct of my life.

But like one walking alone and in the dark, I resolved to proceed so slowly and with such circumspection, that if I did not advance far, I would at least guard against falling. I did not even choose to dismiss summarily any of the opinions that had crept into my belief without having been introduced by reason, but first of all took sufficient time carefully to satisfy myself of the general nature of the task I was setting myself, and ascertain the true method by which to arrive at the knowledge of whatever lay within the compass of my powers.

Among the branches of philosophy, I had, at an earlier period, given some attention to logic, and among those of the mathematics to geometrical analysis and algebra,--three arts or sciences which ought, as I conceived, to contribute something to my design. But, on examination, I found that, as for logic, its syllogisms and the majority of its other precepts are of avail--rather in the communication of what we already know, or even as the art of Lully, in speaking without judgment of things of which we are ignorant, than in the investigation of the unknown; and although this science contains indeed a number of correct and very excellent precepts, there are, nevertheless, so many others, and these either injurious or superfluous, mingled with the former, that it is almost quite as difficult to effect a severance of the true from the false as it is to extract a Diana or a Minerva from a rough block of marble. Then as to the analysis of the ancients and the algebra of the moderns, besides that they embrace only matters highly abstract, and, to appearance, of no use, the former is so exclusively restricted to the consideration of figures, that it can exercise the understanding only on condition of greatly fatiguing the imagination; and, in the latter, there is so complete a subjection to certain rules and formulas, that there results an art full of confusion and obscurity calculated to embarrass, instead of a science fitted to cultivate the mind. By these considerations I was induced to seek some other method which would comprise the advantages of the three and be exempt from their defects. And as a multitude of laws often only hampers justice, so that a state is best governed when, with few laws, these are rigidly administered; in like manner, instead of the great number of precepts of which logic is composed, I believed that the four following would prove perfectly sufficient for me, provided I took the firm and unwavering resolution never in a single instance to fail in observing them.

The first was never to accept anything for true which I did not clearly know to be such; that is to say, carefully to avoid precipitancy and prejudice, and to comprise nothing more in my judgement than what was presented to my mind so clearly and distinctly as to exclude all ground of doubt.

The second, to divide each of the difficulties under examination into as many parts as possible, and as might be necessary for its adequate solution.

The third, to conduct my thoughts in such order that, by commencing with objects the simplest and easiest to know, I might ascend by little and little, and, as it were, step by step, to the knowledge of the more complex; assigning in thought a certain order even to those objects which in their own nature do not stand in a relation of antecedence and sequence.

And the last, in every case to make enumerations so complete, and reviews so general, that I might be assured that nothing was omitted.

The long chains of simple and easy reasonings by means of which geometers are accustomed to reach the conclusions of their most difficult demonstrations, had led me to imagine that all things, to the knowledge of which man is competent, are mutually connected in the same way, and that there is nothing so far removed from us as to be beyond our reach, or so hidden that we cannot discover it, provided only we abstain from accepting the false for the true, and always preserve in our thoughts the order necessary for the deduction of one truth from another. And I had little difficulty in determining the objects with which it was necessary to commence, for I was already persuaded that it must be with the simplest and easiest

to know, and, considering that of all those who have hitherto sought truth in the sciences, the mathematicians alone have been able to find any demonstrations, that is, any certain and evident reasons, I did not doubt but that such must have been the rule of their investigations. I resolved to commence, therefore, with the examination of the simplest objects, not anticipating, however, from this any other advantage than that to be found in accustoming my mind to the love and nourishment of truth, and to a distaste for all such reasonings as were unsound. But I had no intention on that account of attempting to master all the particular sciences commonly denominated mathematics: but observing that, however different their objects, they all agree in considering only the various relations or proportions subsisting among those objects, I thought it best for my purpose to consider these proportions in the most general form possible, without referring them to any objects in particular, except such as would most facilitate the knowledge of them, and without by any means restricting them to these, that afterwards I might thus be the better able to apply them to every other class of objects to which they are legitimately applicable. Perceiving further, that in order to understand these relations I should sometimes have to consider them one by one and sometimes only to bear them in mind, or embrace them in the aggregate, I thought that, in order the better to consider them individually, I should view them as subsisting between straight lines, than which I could find no objects more simple, or capable of being more distinctly represented to my imagination and senses; and on the other hand, that in order to retain them in the memory or embrace an aggregate of many, I should express them by certain characters the briefest possible. In this way I believed that I could borrow all that was best both in geometrical analysis and in algebra, and correct all the defects of the one by help of the other.

And, in point of fact, the accurate observance of these few precepts gave me, I take the liberty of saying, such ease in unraveling all the questions embraced in these two sciences, that in the two or three months I devoted to their examination, not only did I reach solutions of questions I had formerly deemed exceedingly difficult but even as regards questions of the solution of which I continued ignorant, I was enabled, as it appeared to me, to determine the means whereby, and the extent to which a solution was possible; results attributable to the circumstance that I commenced with the simplest and most general truths, and that thus each truth discovered was a rule available in the discovery of subsequent ones Nor in this perhaps shall I appear too vain, if it be considered that, as the truth on any particular point is one whoever apprehends the truth, knows all that on that point can be known. The child, for example, who has been instructed in the elements of arithmetic, and has made a particular addition, according to rule, may be assured that he has found, with respect to the sum of the numbers before him, and that in this instance is within the reach of human genius. Now, in conclusion, the method which teaches adherence to the true order, and an exact enumeration of all the conditions of the thing sought includes all that gives certitude to the rules of arithmetic.

But the chief ground of my satisfaction with thus method, was the assurance I had of thereby exercising my reason in all matters, if not with absolute perfection, at least with the greatest attainable by me: besides, I was conscious that by its use my mind was becoming gradually habituated to clearer and more distinct conceptions of its objects; and I hoped also, from not having restricted this method to any particular matter, to apply it to the difficulties of the other sciences, with not less success than to those of algebra. I should not, however, on this account have ventured at once on the examination of all the difficulties of the sciences which presented themselves to me, for this would have been contrary to the order prescribed in the method, but observing that the knowledge of such is dependent on principles borrowed from philosophy, in which I found nothing certain, I thought it necessary first of all to endeavor to establish its principles. And because I observed, besides, that an inquiry of this kind was of all others of the greatest moment, and one in which precipitancy and anticipation in judgment were most to be dreaded, I thought that I ought not to approach it till I had reached a more mature age (being at that time but twenty-three), and had first of all employed much of my time in preparation for the work, as well by eradicating from my mind all the erroneous opinions I had up to that moment accepted, as by amassing variety of experience to afford materials for my reasonings, and by continually exercising myself in my chosen method with a view to increased skill in its application.

Philosophical Concepts in Physics

By James T. Cushing

In this chapter we discuss a few of the classic views of science that were popular from the time of the Renaissance until the early part of the twentieth century and then indicate some later changes in the conception of science. We return to a more complete overview of the current status of the philosophy of science as a retrospective in Chapter 25.

3.1 Origins of Scientific Method

In the previous chapter we used Bacon as an example of a proponent of what developed into one important aspect of the modern scientific method. We also referred to Descartes as the father of modern philosophy and scientific reasoning. Galileo is often credited as being the first working scientist to apply modern scientific method in his investigations. (Chapter 6 will discuss the scientific writings and research of Galileo.) Although it is simplest for purposes of exposition to focus on the works of specific individuals such as Bacon, Descartes or Galileo to illustrate the rise of modern scientific thought and practice, these seventeenth-century thinkers were not the first to break with Aristotelian tradition. They did have predecessors. For instance, in the thirteenth century an experimental dimension for science was already advocated by the English Franciscan friar Roger Bacon (c. 1220–1292). And, as we shall see in more detail in Chapter 6, some of the fourteenth-century Ockhamists in Paris applied mathematical methods to the problem of motion and obtained results that contributed to the foundations of modern mechanics and of calculus. (This group was named after William of Ockham (c. 1285–1347/1349), famous for his often-quoted razor (or criterion for assessing an explanation), sometimes paraphrased as 'That is best that is simplest and works.') John Buridan (1300–1358) formulated a concept of impetus and treated the effect of gravity on a falling body in terms of what we would today call uniformly accelerated motion. Nicholas Oresme (1325–1382) anticipated some of the elements of analytic geometry by employing coordinates to represent arbitrary functions and reasoned to a distance-time formula for uniformly accelerated motion. Leonardo da Vinci (1452–1519) knew of the work of this Parisian School and his own investigations in mechanics were influenced by it. He stressed the importance of using mathematics in science, the need for observation and experiment, and the reciprocal roles played by (in modern terminology) both induction and deduction in discovering the laws of nature. The point is that, even though we often present new ideas as represented by one person, it is necessary to realize that the roots of revolutionary insights usually go far into the past and have involved histories. Here we can give only a brief outline of their origins.

Underlying the work of Bacon, of Galileo and of other proponents of the 'scientific method' is a belief in the objective validity of science. This means that nature really exists outside of and independently of us, that there are laws of nature that function without regard for our expectations, and, moreover, that we are capable of discovering those laws. This is an act of faith on the part of the scientist. Science itself cannot prove the correctness or truth of this basic assumption of an external world governed by knowable laws. Perhaps this nondemonstrable element of belief at the base of the scientific enterprise strikes you as alien to some preconceived notions you may have about the nature of science. If that is the case, then the present chapter could be a bit disquieting.

So, before turning in succeeding chapters to our study of concrete physical laws, we make a few remarks on the nature of the scientific enterprise.

3.2 A Popular View of Science

A current popular conception of science, that bears some similarity to the type of program proposed by Francis Bacon, can be set down in outline form as:

a. observation
b. hypothesis
c. prediction
d. confirmation

This simple model of science and how it operates depicts the process as a purely objective one in which careful observations are made, or controlled experiments performed, after which rules or laws are extracted from these empirical data and general hypotheses are formed. Next, these general hypotheses or conjectured theories are employed to predict new observational consequences. These predictions are used to confront reality and, if the predictions are verified, then the theory has been confirmed. By this process science amasses more and more knowledge about the physical world and brings us closer to the truth.

Let us now examine this model a bit more critically. Certainly nearly all physical scientists have agreed that one must begin with the data of sense experience. Not only men like Bacon, Galileo and Newton were of this persuasion. Harvey, in his revolutionary and pioneering treatise on the circulation of the blood in the human body, urged, as Aristotle had, that we proceed from things more known to those less known. This approach is elegantly and concisely summarized in his statement 'that the facts cognizable by the senses wait upon no opinions, and that the works of nature bow to no antiquity' But what are we to do with these facts and data?

3.3 Hume and Mill on Induction

Previously we termed induction the process of abstracting general laws or principles from specific observations. Can one ever know with certainty that a law gained from induction is correct? David Hume, the eighteenth-century British philosopher, historian, economist and essayist, held that philosophy is basically an inductive, experimental science of human nature. In his An Enquiry Concerning Human Understanding (1758), he discusses the notion of causality, the validity of induction and the necessity of sense data preceding ideas. Here we simply summarize the quotation from Hume's work given in Section 3.A. In analyzing the concepts of cause and effect, Hume points out that what we actually observe is only one event following another, rather than a necessary connection between them. There is no direct observation of a cause (or connection) for the supposed effect. Since what we see are events conjoined (or associated) but never connected, the terms cause and effect are words without operational meaning. If we know just one instance of two events being conjoined (one following another), then we do not state a general law governing their connection in terms of cause and effect. However, if we repeatedly see the same pair of events conjoined, as a body initially at rest being struck by another moving object so that after the collision the first body is also in motion, we do refer to cause and effect. (In this example, the collision of the moving object with the body at rest would be termed the cause for the effect of that body being set into motion.) That is, from the occurrence of constant conjunction, we infer a necessary connection. However, the difference between one such pair of events and many is that the mind is carried by habit to feel the existence of a necessary connection (in our thought).

In essence, Hume asked what reason we have to suppose that future observations would resemble past ones. He challenged placing unlimited trust in universal, predictive laws induced from observed facts. As a simple example, from the fact that the sun has risen every day in the past since the beginning of recorded history, we might induce the 'law' that the sun necessarily rises every day without exception. Can we, however, be absolutely certain that the sun will rise tomorrow? No. There is no way around the provisional character of physical laws gained by induction. Hume saw the need for induction, but felt its validity to be unprovable.

The later British philosopher, economist and ethical theorist John Mill also wrote extensively on the question of induction. Mill was a child prodigy educated by his father so that he read Greek by age eight, when he started Latin, studied Euclid and then Aristotle at age twelve. He was not only a philosopher, but a political activist as well. In his System of Logic (1843), Mill discusses induction and the assumed uniformity of the laws of nature. He argues that the main problem for the science of logic is a justification

of the process of induction that is the operation of discovering and proving general propositions. Mill sees the question to be one of discovering the conditions under which an induction is logically legitimate. Involved in all inductions is the assumption that the course of nature is uniform or that the universe is governed by general laws. However, this is itself an example of induction since we assume it to be true because it has always turned out to be so in individual instances. This uniformity of the course of nature is the premise (or first principle) that can be used to justify induction. Unfortunately, neither Mill nor anyone else has proven the correctness of this premise. It remains an assumption, albeit one that few would dispute and without which we cannot do science.

3.4 Popper on Observation and Hypothesis

You may feel that a concern for the nature of the cause-effect relationship, the validity of induction and the status of physical laws is just philosophical nit-picking that is best left for some idle Saturday afternoon and that we can get on with the serious business of science without being concerned about such questions. But even if we gloss over these issues, consider the relation between observation and hypothesis. How independent are the operations of gathering data and of discovering theories? The fact-gathering necessary for science is rarely random. As Charles Darwin (1809–1882), whose On the Origin of Species (1859) laid the foundation of the theory of evolution, pointed out, no one can be a good observer unless he is an active theorizer. That is, the discovery of new facts and the evolution of a theory are reciprocal processes. The experiments a scientist undertakes do not yield immediate and obvious sense data but rather those collected with considerable difficulty. She must have a theory to guide her in this search, else she would not know which of the myriad phenomena in nature to examine. As we shall see in our study of physics, those phenomena that one age sees as crucial in light of a then-current theory often appear less important at a later time when a different theory has become more popular.

Perhaps you object that these are still mere details of the relationship between observation and hypothesis. One way or another, scientists do arrive at candidates for physical laws and then from these deduce new consequences that can be looked for. What do we conclude when a theory predicts a result that is found to correspond with our observations? Has the theory been verified? No, we simply know that it has not yet been refuted. Sometimes we put a more positive gloss on this by speaking of corroboration or of confirmation. That is, if a theory makes a prediction that is definitely contradicted by sufficiently careful observation or experimentation, then the theory is necessarily logically incorrect. It has been refuted. A given theory (p) implies or makes a specific prediction (q). If the prediction turns out to be false (~q) as indicated by further analysis or by experimental results, then it necessarily follows that the theory is false (~p). (Unfortunately, it is not always a simple matter to know when the data are sufficiently good to refute an hypothesis, as we shall see in Chapters 4 and 5 when we discuss stellar parallax.) However, if the prediction is correct (q), we cannot be certain that the theory (p) is correct.

One recent British philosopher of science, Sir Karl Popper (1902–1994), stressed that the hallmark of a scientific theory is not that it can be verified but rather that it is capable of being refuted or falsified. According to this criterion, a theory can qualify as scientific only if it makes specific predictions that can be subjected to comparison with the real world to find out whether or not these predictions are indeed true in physical reality. In his Conjectures and Refutations, an excerpt of which is given at the end of this chapter, Popper asks what the criterion is by which one can distinguish a scientific theory from a nonscientific one. As a young man (around 1919), he contrasted

Karl Marx's (1818–1883) theory of history, Sigmund Freud's (1856–1939) psychoanalysis and Alfred Adler's (1870–1937) individual psychology with Einstein's general theory of relativity. Since the former three all had great explanatory power in that whatever happened could always be interpreted after the fact as confirming those theories, Popper concluded that verifications alone are not of central importance since in these cases they are so easily come by. In the case of general relativity, or of any truly scientific theory, definite predictions are made that are subsequently tested by experiment. There is no guarantee ahead of time that the observational facts will not disagree with these predictions. Such theories are refutable. Only predictions that have this element of risk (in that they might be refuted) should count as meaningful supportive evidence for scientific theories. A good scientific theory prohibits certain outcomes from occurring in nature and a stringent test of such a theory is an attempt to falsify or refute it by actually observing those prohibited results. Hence, for Popper, the hallmark of scientific theories is that they are (in principle) refutable or falsifiable. (This is not the same as saying that they are in fact constantly refuted. A successful scientific theory survives many serious attempts to refute it.) This is similar to the position of the American philosopher Willard Quine(1908–). (Recall the quotations for Part I (The scientific enterprise).)

3.5 Justification for Hypotheses

Therefore, as long as a scientific theory holds up under comparison with observations, it may be correct. We can never be certain that a theory is correct, only that refuted theories are incorrect. Still, scientists do use theories to make predictions and, when these predictions accord with nature, then scientists continue to work with such theories. This model of postulating a theory and then making specific predictions with it is termed the hypothetico-deductive method. Although that method does not prove a theory as necessarily correct, it does give us a warrant for accepting the theory (always provisionally of course). Retroduction, as we define it here, signifies arguing for the plausibility of a theory on the basis of its successful specific predictions. From our modern perspective, we can recognize that this is the method that Newton advocates in his *Rules of Reasoning in Philosophy*.

> In experimental philosophy we are to look upon propositions inferred by general induction from phenomena as accurately or very nearly true, notwithstanding any contrary hypotheses that may be imagined, till such time as other phenomena occur, by which they may either be made more accurate, or liable to exceptions.[2]

So far we have seen three types of warrants for the axioms or postulates from which we then deduce logically certain implications or predictions:

i. *axiomatic*-Here the axiom is claimed to be self-evident, obvious or immediate. As we saw in Chapter 2, Aristotle often arrived at his general cosmological principles in this fashion after a rather cursory examination of the data of experience. Descartes attempted to base the science of mechanics on such first principles seen to be true in their own right, once properly understood.
ii. *inductive*-Here generalizations are made from similarities perceived in a large group of particular events or observations. This is the method proposed by Bacon and later espoused by Hume and by Mill.
iii. *retroductive*-Here we go from consequences back to hypotheses as in the hypothetico-deductive method above.

Let us return now to the falsification process that Popper feels is essential for the operation of science.

Is this process quite so simple and certain? No, since observational data alone are never sufficient to force scientists to abandon a model or theory that has previously been successful. Such theories are often modified or reinterpreted. As a case in point, Planck was involved in a long and vitriolic controversy with several distinguished colleagues over the interpretation and relation of some of the laws of heat and of mechanics. Even though he had sound arguments on his side, it was many years before his point of view was generally accepted. In his Scientific Autobiography he states:

> This experience gave me also an opportunity to learn a fact—a remarkable one in my opinion: a new scientific truth does not triumph by convincing its opponents and making them see the light, but rather because its opponents eventually die, and a new generation grows up that is familiar with it.[3]

For example, we shall see that observation alone did not decide the issue between Claudius Ptolemy's (*fl.* 127–145 A.D.) model of the universe that placed the earth at its center and Nicholas Copernicus' (1473–1543) model that placed the sun at its center. If not observation alone, then what are the other criteria by which scientists decide among rival theories? A certain economy of explanation, a symmetry of form and a beauty of formulation have always been important factors. These are aesthetic matters that cannot be quantified and are not agreed upon universally. For some, discovering beauty and a sense of symmetry is the end of science and the reason for doing it. We return to these questions several times throughout this book.

3.6 Scientific Knowledge and Truth

Even though the absolute objectivity and certitude of science have now been opened to some doubt, you may still hold that nevertheless science both seeks and finds truth. Is this obvious, though? Since we can never be certain that a given model or theory of the physical world is correct, can we say that such theories are true? As we emphasized previously, there is an essential asymmetry between the confirmation and the refutation of a theory or of a proposition. Refutation is a logically certain and valid inference in that, once the prediction (or consequent) has been denied, then the theory (or antecedent) must also be denied. However, there is no logically valid means of establishing the truth of a theory from the agreement of its predictions with observed physical phenomena. (An exception to this would occur if one were able to show that there existed only a finite number of logically possible explanations and all but one were refuted, say by observation. However, this is rare in complex, realistic situations.) Even if a theory were true (in the sense that it did actually embody and uniquely account for an absolutely and universally valid law of nature), we would have no way to know that it was true. Are not such theories rather constructs of the human mind by which we represent nature? One can take scientific theories to be just a means of simplifying and codifying the observed facts about nature, and not an explanation of them.

In his 1941 lecture The Meaning and Limits of Exact Science, Planck discussed the task of science as introducing order and regularity into the multitude of diverse sense experiences. He observed that, even though the individual data of sense experience have great solidity, they are of limited significance. We feel compelled to postulate the existence of a real world that underlies and unifies these diverse surface phenomena. This belief constitutes an irreducible, nondemonstrable element of science, but one we feel unable to dispense with. According to Planck, science studies the phenomenological world in order to construct a scientific world picture of our experiences and from this to gain an understanding of what he calls the real world of objects, the ultimate, metaphysical reality that can never be completely known. We approach the metaphysically real world through improvement of our phenomenological world picture. Planck's real metaphysical world of objects is similar to Plato's world of forms or essences that we mentioned in Chapter 1.

If we take a brief retrospective glance at the evolution of the concept and status of scientific knowledge as we have presented it in these first three chapters, we can associate the major developments with the following sequence of philosophers.

Plato, Aristotle $\rightarrow$ Bacon, Descartes $\rightarrow$ Galileo, Newton $\rightarrow$ Hume, Mill $\rightarrow$ Popper, Quine

What began as a quest for certain knowledge and understanding of the phenomena of nature, based either on deduction from self-evident first principles (Aristotle, Descartes) or on careful induction to lead to unassailable general laws (Bacon), has today been abandoned as unattainable. One modern philosopher has referred to this goal of attainable certain knowledge as the Bacon-Descartes ideal.[4] The basic problem with that program is that there is no logically valid way to arrive at first principles (or general laws), either on the basis of intuition, of induction or of a combination of the two. A content-increasing logic (or a logic of induction) does not exist. Deductive logic merely allows us to search out those statements already contained (implicitly) in the premises. The conclusion contains no more facts or information than the premises (even though it may be more useful or easily recognizable there than in the premises). The trend of these philosophical developments has been a decrease in the certainty we are able to claim for scientific knowledge. We can only hope to refute candidates for scientific theories, not to prove them as true. From this perspective our successful, accepted scientific theories are consistent, but not necessarily 'true', stories about the way the world might 'really' be. In the next chapter we begin to examine how the history of actual scientific practice leads us to such an evaluation of the status of scientific theories.

3.7 Hume, Mill and Popper on Scientific Knowledge

In his An Enquiry Concerning Human Understanding Hume discussed the concepts of cause and effect.

> We have sought in vain for an idea of power or necessary connexion in all the sources from which we could suppose it to be derived. It appears that, in single instances of the operation of bodies, we never can, by our utmost scrutiny, discover anything but one event following another, without being able to comprehend any force or power by which the cause operates, or any connection between it and its supposed effect. All events seem entirely loose and separate. One event follows another; but we never can observe any tie between them. They seem conjoined, but never connected. And as we can have no idea of any thing which never appeared to our outward sense or inward sentiment, the necessary conclusion seems to be that we have no idea of connection or power at all, and that these words are absolutely without any meaning, when employed either in philosophical reasonings or common life.

> * * *

> Even after one instance or experiment where we have observed a particular event to follow upon another, we are not entitled to form a general rule, or foretell what will happen in like cases; it being justly esteemed an unpardonable temerity to judge of the whole course of nature from one single experiment, however accurate or certain. But when one particular species of event has always, in all instances, been conjoined with another, we make no longer any scruple of foretelling one upon the appearance of the other, and of employing that reasoning, which can alone assure us of any matter of fact or existence. We then call the one object, Cause; the other, Effect. We suppose that there is some connexion between them; some power in the one, by which it infallibly produces the other, and operates with the greatest certainty and strongest necessity.

> It appears, then, that this idea of a necessary connexion among events arises from a number of similar instances which occur of the constant conjunction of these events; nor can that idea ever be suggested by any one of these instances, surveyed in all possible lights and positions. But there is nothing in a number of instances, different from every single instance, which is supposed to be exactly similar; except only, that after a repetition of similar instances, the mind is carried by habit, upon the appearance of one event, to expect its usual attendant, and to believe that it will exist. This connection, therefore, which we **feel** in the mind, this customary transition of the imagination from one object to its usual attendant, is the sentiment or impression from which we form the idea of power or necessary connection. Nothing farther is in the case. Contemplate the subject on all sides; you will never find any other origin of that idea. This is the sole difference between one instance, from which we can never receive the idea of connection, and a number of similar instances, by which it is suggested.

> * * *

[T]herefore, we may define a cause to be *an object, followed by another, and where all the objects similar to the first are followed by objects similar to the second.* Or in other words *where, if the first object had not been, the second never had existed.* The appearance of a cause always conveys the mind, by a customary transition, to the idea of the effect. Of this also we have experience. We may, therefore, suitably to this experience, form another definition of cause, and call it, an object *followed by another, and whose appearance always conveys the thought to that other.*[5]

In A System of Logic Mill turned to the problem of induction.

What Induction is, therefore, and what conditions render it legitimate, cannot but be deemed the main question of the science of logic—the question which includes all others.

* * *

For the purposes of the present inquiry, Induction may be defined: the operation of discovering and proving general propositions.

* * *

Induction properly so called, as distinguished from those mental operations, sometimes, though improperly, designated by the name, which I have attempted in the preceding chapter to characterize, may, then, be summarily defined as generalization from experience. It consists in inferring from some individual instances in which a phenomenon is observed to occur, that it occurs in all instances of a certain class; namely, in all which resemble the former, in what are regarded as the material circumstances.

In what way the material circumstances are to be distinguished from those which are immaterial, or why some of the circumstances are material and others not so, we are not yet ready to point out. We must first observe that there is a principle implied in the very statement of what Induction is; an assumption with regard to the course of nature and the order of the universe: namely, that there are such things in nature as parallel cases; that what happens once, will, under a sufficient degree of similarity of circumstances, happen again, and not only again, but always. This, I say, is an assumption involved in every case of induction. And, if we consult the actual course of nature, we find that the assumption is warranted; the fact is so. The universe, we find, is so constituted, that whatever is true in one case, is true in all cases of a certain description; the only difficulty is, to find what description.

This universal fact, which is our warrant for all inferences from experience, has been described by different philosophers in different forms of language: that the course of nature is uniform; that the universe is governed by general laws; and the like.

* * *

[T]he principle which we are now considering, that of the uniformity of the course of nature, will appear as the ultimate major premise of all inductions; and will, therefore, stand to all inductions in the relation in which, as has been shown at so much length, the major proposition of a syllogism always stands to the conclusion; not contributing at all to prove it, but being a necessary condition of its being proved;

since no conclusion is proved for which there cannot be found a true major premiss.[6]

In his Conjectures and Refutations Popper addressed the concept of falsification.

I therefore decided to do what I have never done before: to give you a report on my own work in the philosophy of science, since the autumn of 1919 when I first began to grapple with the problem, 'When should a theory be ranked as scientific?' or Is there a criterion for the scientific character or status of a theory?'

* * *

I knew, of course, the most widely accepted answer to my problem: that science is distinguished from pseudo-science-or from 'metaphysics'—by its empirical method, which is essentially inductive, proceeding from observation or experiment. But this did not satisfy me. On the contrary, I often formulated my problem as one of distinguishing between a genuinely empirical method and a non-empirical or even a pseudo-empirical method—that is to say, a method which, although it appeals to observation and experiment, nevertheless does not come up to scientific standards. The latter method may be exemplified by astrology, with its stupendous mass of empirical evidence based on observation—on horoscopes and on biographies.

But as it was not the example of astrology which led me to my problem I should perhaps briefly describe the atmosphere in which my problem arose and the examples by which it was stimulated. After the collapse of the Austrian Empire there had been a revolution in Austria: the air was full of revolutionary slogans and ideas,

and new and often wild theories. Among the theories which interested me Einstein's theory of relativity was no doubt by far the most important. Three others were Marx's theory of history, Freud's psycho-analysis, and Alfred Adler's so-called 'individual psychology'.

* * *

I found that those of my friends who were admirers of Marx, Freud, and Adler were impressed by a number of points common to these theories, and especially by their apparent explanatory power. These theories appeared to be able to explain practically everything that happened within the fields to which they referred. The study of any of them seemed to have the effect of an intellectual conversion or revelation, opening your eyes to a new truth hidden from those not yet initiated. Once your eyes were thus opened you saw confirming instances everywhere: the world was full of verifications of the theory. Whatever happened always confirmed it. Thus its truth appeared manifest; and the unbelievers were clearly people who did not want to see the manifest truth; who refused to see it, either because it was against their class interest, or because of their repressions which were still 'unanalyzed' and crying aloud for treatment.

* * *

With Einstein's theory the situation was strikingly different. Take one typical instance—Einstein's prediction, just then confirmed by the findings of Eddington's expedition. Einstein's gravitational theory had led to the result that light must be attracted by heavy bodies (such as the sun), precisely as material bodies were attracted. As a consequence it could be calculated that light from a distant fixed star whose apparent position was close to the sun would reach the earth from such a direction that the star would seem to be slightly shifted away from the sun; or, in other words, that the stars close to the sun would look as if they had moved a little away from the sun, and from one another.

* * *

Now the impressive thing about this case is the risk involved in a prediction of this kind. If observation shows that the predicted effect is definitely absent, then the theory is simply refuted. The theory is incompatible with certain possible results of observation—in fact with results which everybody before Einstein would have expected. This is quite different from the situation I have previously described, when it turned out that the theories in question were compatible with the most divergent human behaviour, so that it was practically impossible to describe any human

behaviour that might not be claimed to be a verification of these theories.

* * *

One can sum up all this by saying that the criterion of the scientific status of a theory is its falsifiability, or refutability, or testability.[7]

Further Reading

Ralph Blake et al.'s Theories of Scientific Method gives a nice overview of this subject from the Renaissance through the nineteenth century. Karl Popper's Conjectures and Refutations contains essays and lectures on the theme 'that we can learn from our mistakes' (via falsifiability) and sets this thesis against a broad historical background of philosophical positions on scientific knowledge. Baruch Brody's Readings in the Philosophy of Science is a collection of papers, by prominent modern philosophers of science, on scientific explanation and prediction, on the structure and function of scientific theories and on the confirmation of scientific hypotheses.

The Structure of Scientific Revolutions

By T. S. Kuhn

VII. Crisis and the Emergence of Scientific Theories

All the discoveries considered in Section VI were causes of or contributors to paradigm change. Furthermore, the changes in which these discoveries were implicated were all destructive as well as constructive. After the discovery had been assimilated, scientists were able to account for a wider range of natural phenomena or to account with greater precision for some of those previously known. But that gain was achieved only by discarding some previously standard beliefs or procedures and, simultaneously, by replacing those components of the previous paradigm with others. Shifts of this sort are, I have argued, associated with all discoveries achieved through normal science, excepting only the unsurprising ones that had been anticipated in all but their details. Discoveries are not, however, the only sources of these destructive-constructive paradigm changes. In this section we shall begin to consider the similar, but usually far larger, shifts that result from the invention of new theories.

Having argued already that in the sciences fact and theory, discovery and invention, are not categorically and permanently distinct, we can anticipate overlap between this section and the last. (The impossible suggestion that Priestley first discovered oxygen and Lavoisier then invented it has its attractions. Oxygen has already been encountered as discovery; we shall shortly meet it again as invention.) In taking up the emergence of new theories we shall inevitably extend our understanding of discovery as well. Still, overlap is not identity. The sorts of discoveries considered in the last section were not, at least singly, responsible for such paradigm shifts as the Copernican, Newtonian, chemical, and Einsteinian revolutions. Nor were they responsible for the somewhat smaller, because more exclusively professional, changes in paradigm produced by the wave theory of light, the dynamical theory of heat, or Maxwell's electromagnetic theory. How can theories like these arise from normalscience, an activity even less directed to their pursuit than to that of discoveries?

If awareness of anomaly plays a role in the emergence of new sorts of phenomena, it should surprise no one that a similar but more profound awareness is prerequisite to all acceptable changes of theory. On this point historical evidence is, I think, entirely unequivocal. The state of Ptolemaic astronomy was a scandal before Copernicus' announcement[1] Galileo's

---

[1] A. R. Hall, *The Scientific Revolution, 1500–1800* (London, 1954), p. 16.

contributions to the study of motion depended closely upon difficulties discovered in Aristotle's theory by scholastic critics.[2] Newton's new theory of light and color originated in the discovery that none of the existing pre-paradigm theories would account for the length of the spectrum, and the wave theory that replaced Newton's was announced in the midst of growing concern about anomalies in the relation of diffraction and polarization effects to Newton's theory.[3] Thermodynamics was born from the collision of two existing nineteenth-century physical theories, and quantum mechanics from a variety of difficulties surrounding black-body radiation, specific heats, and the photoelectric effect.[4] Furthermore, in all these cases except that of Newton the awareness of anomaly had lasted so long and penetrated so deep that one can appropriately describe the fields affected by it as in a state of growing crisis. Because it demands large-scale paradigm destruction and major shifts in the problems and techniques of normal science, the emergence of new theories is generally preceded by a period of pronounced professional in-security. As one might expect, that insecurity is generated by the persistent failure of the puzzles of normal science to come out as

they should. Failure of existing rules is the prelude to a search for new ones.

Look first at a particularly famous case of paradigm change, the emergence of Copemican astronomy. When its predecessor, the Ptolemaic system, was first developed during the last two centuries before Christ and the first two after, it was admirably successful in predicting the changing positions of both stars and planets. No other ancient system had performed so well; for the stars, Ptolemaic astronomy is still widely used today as an engineering approximation; for the planets, Ptolemy's predictions were as good as Copernicus'. But to be admirably successful is never, for a scientific theory, to be completely successful. With respect both to planetary position and to precession of the equinoxes, predictions made with Ptolemy's system never quite conformed with the best available observations. Further reduction of those minor discrepancies constituted many of the principal problems of normal astronomical research for many of Ptolemy's successors, just as a similar attempt to bring celestial observation and Newtonian theory together provided normal research problems for Newton's eighteenth-century successors. For some time astronomers had every reason to suppose that these attempts would be as successful as those that had led to Ptolemy's system. Given a particular discrepancy, astronomers were invariably able to eliminate it by making some particular adjustment in Ptolemy's system of compounded circles. But as time went on, a man looking at the net result of the normal research effort of many astronomers could observe that astronomy's complexity was increasing far more rapidly than its accuracy and that a discrepancy corrected in one place was likely to show up in another.[5]

Because the astronomical tradition was repeatedly interrupted from outside and because, in the absence of printing, communication between astronomers

---

[2] Marshall Clagett, *The Science of Mechanics in the Middle Ages* (Madison, Wis., 1959), Parts II-III. A. Koyre displays a number of medieval elements in Galileo's thought in his *Etudes Galileennes* (Paris, 1939), particularly Vol. I.

[3] For Newton, see T. S. Kuhn, "Newton's Optical Papers," in *Isaac Newton's Papers and Letters in Natural Philosophy*, ed. I. B. Cohen (Cambridge, Mass., 1958), pp. 27–45. For the prelude to the wave theory, see E. T. Whittaker, *A History of the Theories of Aether and Electricity*, I (2d ed.; London, 1951), 94–109; and W. Whewell, *History of the Inductive Sciences* (rev. ed.; London, 1847), II, 396–466.

[4] For thermodynamics, see Silvanus P. Thompson, *Life of William Thomson Baron Kelvin of Largs* (London, 1910), I, 266–81. For the quantum theory, see Fritz Reichc, *The Quantum Theory*, trans. H. S. Hatfield and II. L. Brose (London, 1922), chaps, i-ii.

[5] J. L. E. Dreyer, *A History of Astronomy from Thales to Kepler* (2d ed.; New York, 1953), chaps, xi-xii.

was restricted, these difficulties were only slowly recognized. But awareness did come. By the thirteenth century Alfonso X could proclaim that if God had consulted him when creating the universe, he would have received good advice. In the sixteenth century, Copernicus' coworker, Domenico da Novara, held that no system so cumbersome and inaccurate as the Ptolemaic had become could possibly be true of nature. And Copernicus himself wrote in the Preface to the *De Revolutionibus* that the astronomical tradition he inherited had finally created only a monster. By the early sixteenth century an increasing number of Europe's best astronomers were recognizing that the astronomical paradigm was failing in application to its own traditional problems. That recognition was prerequisite to Copernicus' rejection of the Ptolemaic paradigm and his search for a new one. His famous preface still provides one of the classic descriptions of a crisis state.[6]

Breakdown of the normal technical puzzle-solving activity is not, of course, the only ingredient of the astronomical crisis that faced Copernicus. An extended treatment would also discuss the social pressure for calendar reform, a pressure that made the puzzle of precession particularly urgent. In addition, a fuller account would consider medieval criticism of Aristotle, the rise of Renaissance Neoplatonism, and other significant historical elements besides. But technical breakdown would still remain the core of the crisis. In a mature science—and astronomy had become that in antiquity—external factors like those cited above are principally significant in determining the timing of breakdown, the ease with which it can be recognized, and the area in which, because it is given particular attention, the breakdown first occurs. Though immensely important, issues of that sort are out of bounds for this essay.

If that much is clear in the case of the Copernican revolution, let us turn from it to a second and rather different example, the crisis that preceded the emergence of Lavoisier's oxygen theory of combustion. In the 1770's many factors combined to generate a crisis in chemistry, and historians are not altogether agreed about either their nature or their relative importance. But two of them are generally accepted as of first-rate significance: the rise of pneumatic chemistry and the question of weight relations. The history of the first begins in the seventeenth century with development of the air pump and its deployment in chemical experimentation. During the following century, using that pump and a number of other pneumatic devices, chemists came increasingly to realize that air must be an active ingredient in chemical reactions. But with a few exceptions—so equivocal that they may not be exceptions at all—chemists continued to believe that air was the only sort of gas. Until 1756, when Joseph Black showed that fixed air (CO_2) was consistently distinguishable from normal air, two samples of gas were thought to be distinct only in their impurities.[7]

After Black's work the investigation of gases proceeded rapidly, most notably in the hands of Cavendish, Priestley, and Scheele, who together developed a number of new techniques capable of distinguishing one sample of gas from another. All these men, from Black through Scheele, believed in the phlogiston theory and often employed it in their design and interpretation of experiments. Scheele actually first produced oxygen by an elaborate chain of experiments designed to dephlogisticate heat. Yet the net result of their experiments was a variety of gas samples and gas properties so elaborate that the phlogiston theory proved increasingly little able to cope with laboratory experience. Though none of these chemists suggested that the theory should be replaced, they were unable to apply it consistently. By

---

[6] T. S. Kuhn, *The Copernican Revolution* (Cambridge, Mass., 1957), pp. 135–43.

[7] J. R. Partington, *A Short History of Chemistry* (2d ed.; London, 1951), pp. 48–51, 73–85, 90–120.

the time Lavoisier began his experiments on airs in the early 1770's, there were almost as many versions of the phlogiston theory as there were pneumatic chemists.[8] That proliferation of versions of a theory is a very usual symptom of crisis. In his preface, Copernicus complained of it as well.

The increasing vagueness and decreasing utility of the phlogiston theory for pneumatic chemistry were not, however, the only source of the crisis that confronted Lavoisier. He was also much concerned to explain the gain in weight that most bodies experience when burned or roasted, and that again is a problem with a long prehistory. At least a few Islamic chemists had known that some metals gain weight when roasted. In the seventeenth century several investigators had concluded from this same fact that a roasted metal takes up some ingredient from the atmosphere. But in the seventeenth century that conclusion seemed unnecessary to most chemists. If chemical reactions could alter the volume, color, and texture of the ingredients, why should they not alter weight as well? Weight was not always taken to be the measure of quantity of matter. Besides, weight-gain on roasting remained an isolated phenomenon. Most natural bodies (e.g., wood) lose weight on roasting as the phlogiston theory was later to say they should.

During the eighteenth century, however, these initially adequate responses to the problem of weight-gain became increasingly difficult to maintain. Partly because the balance was increasingly used as a standard chemical tool and partly because the development of pneumatic chemistry made it possible and desirable to retain the gaseous products of reactions, chemists discovered more and more cases in which weight-gain accompanied roasting. Simultaneously, the gradual assimilation of Newton's gravitational theory led chemists to insist that gain in weight must mean gain in quantity of matter. Those conclusions did not result in rejection of the phlogiston theory, for that theory could be adjusted in many ways. Perhaps phlogiston had negative weight, or perhaps fire particles or something else entered the roasted body as phlogiston left it. There were other explanations besides. But if the problem of weight-gain did not lead to rejection, it did lead to an increasing number of special studies in which this problem bulked large. One of them, "On phlogiston considered as a substance with weight and [analyzed] in terms of the weight changes it produces in bodies with which it unites," was read to the French Academy early in 1772, the year which closed with Lavoisier's delivery of his famous sealed note to the Academy's Secretary. Before that note was written a problem that had been at the edge of the chemist's consciousness for many years had become an outstanding unsolved puzzle.[9] Many different versions of the phlogiston theory were being elaborated to meet it. Like the problems of pneumatic chemistry, those of weight-gain were making it harder and harder to know what the phlogiston theory was. Though still believed and trusted as a working tool, a paradigm of eighteenth-century chemistry was gradually losing its unique status. Increasingly, the research it guided resembled that conducted under the competing schools of the pre-paradigm period, another typical effect of crisis.

Consider now, as a third and final example, the late nineteenth century crisis in physics that prepared the way for the emergence of relativity theory. One root of that crisis can be traced to the late seventeenth century when a number of natural philosophers, most notably Leibniz, criticized Newton's retention of an

---

[8] Though their main, concern is with a slightly later period, much relevant material is scattered throughout J. R. Partington and Douglas McKie's "Historical Studies on the Phlogiston Theory," *Annals of Science,* II (1937), 361–404; III (1938), 1–58, 337–71; and IV (1939), 337–71.

[9] H. Guerlac, *Lavoisier—the Crucial Year* (Ithaca, N.Y., 1961). The entire book documents the evolution and first recognition of a crisis. For a clear statement of the situation with respect to Lavoisier, see p. 35.

updated version of the classic conception of absolute space.[10] They were very nearly, though never quite, able to show that absolute positions and absolute motions were without any function at all in Newton's system; and they did succeed in hinting at the considerable aesthetic appeal a fully relativistic conception of space and motion would later come to display. But their critique was purely logical. Like the early Copernicans who criticized Aristotle's proofs of the earth's stability, they did not dream that transition to a relativistic system could have observational consequences. At no point did they relate their views to any problems that arose when applying Newtonian theory to nature. As a result, their views died with them during the early decades of the eighteenth century to be resurrected only in the last decades of the nineteenth when they had a very different relation to the practice of physics.

The technical problems to which a relativistic philosophy of space was ultimately to be related began to enter normal science with the acceptance of the wave theory of light after about 1815, though they evoked no crisis until the 1890's. If light is wave motion propagated in a mechanical ether governed by Newton's Laws, then both celestial observation and terrestrial experiment become potentially capable of detecting drift through the ether. Of the celestial observations, only those of aberration promised sufficient accuracy to provide relevant information, and the detection of ether-drift by aberration measurements therefore became a recognized problem for normal research. Much special equipment was built to resolve it. That equipment, however, detected no observable drift, and the problem was therefore transferred from the experimentalists and observers to the theoreticians. During the central decades of the century Fresnel, Stokes, and others devised numerous articulations of the ether theory designed to explain the failure to observe drift. Each of these articulations assumed

that a moving body drags some fraction of the ether with it. And each was sufficiently successful to explain the negative results not only of celestial observation but also of terrestrial experimentation, including the famous experiment of Michelson and Morley.[11] There was still no conflict excepting that between the various articulations. In the absence of relevant experimental techniques, that conflict never became acute.

The situation changed again only with the gradual acceptance of Maxwell's electromagnetic theory in the last two decades of the nineteenth century. Maxwell himself was a Newtonian who believed that light and electromagnetism in general were due to variable displacements of the particles of a mechanical ether. His earliest versions of a theory for electricity and magnetism made direct use of hypothetical properties with which he endowed this medium. These were dropped from his final version, but he still believed his electromagnetic theory compatible with some articulation of the Newtonian mechanical view.[12] Developing a suitable articulation was a challenge for him and his successors. In practice, however, as has happened again and again in scientific development, the required articulation proved immensely difficult to produce. Just as Copernicus' astronomical proposal, despite the optimism of its author, created an increasing crisis for existing theories of motion, so Maxwell's theory, despite its Newtonian origin, ultimately produced a crisis for the paradigm from which it had sprung.[13] Furthermore, the locus at which that crisis became most acute was provided by

---

[10] Max Jammer, *Concepts of Space: The History of Theories of Space in Physics.* (Cambridge, Mass., 1954), pp. 114–24.

[11] Joseph Larmor, *Aether and Matter … Including a Discussion of the Influence of the Earth's Motion on Optical Phenomena* (Cambridge, 1900), pp. 6–20, 320–22.

[12] R. T. Glazebrook, *James Clerk Maxwell and Modern Physics* (London, 1896), chap. ix. For Maxwell's final attitude, see his own book, A *Treatise on Electricity and Magnetism* (3d ed.; Oxford, 1892), p. 470.

[13] For astronomy's role in the development of mechanics, see Kuhn, *op. cit.,* chap. vii.

the problems we have just been considering, those of motion with respect to the ether.

Maxwell's discussion of the electromagnetic behavior of bodies in motion had made no reference to ether drag, and it proved very difficult to introduce such drag into his theory. As a result, a whole series of earlier observations designed to detect drift through the ether became anomalous. The years after 1890 therefore witnessed a long series of attempts, both experimental and theoretical, to detect motion with respect to the ether and to work ether drag into Maxwell's theory. The former were uniformly unsuccessful, though some analysts thought their results equivocal. The latter produced a number of promising starts, particularly those of Lorentz and Fitzgerald, but they also disclosed still other puzzles and finally resulted in just that proliferation of competing theories that we have previously found to be the concomitant of crisis.[14] It is against that historical setting that Einstein's special theory of relativity emerged in 1905.

These three examples are almost entirely typical. In each case a novel theory emerged only after a pronounced failure in the normal problem-solving activity. Furthermore, except for the case of Copernicus in which factors external to science played a particularly large role, that breakdown and the proliferation of theories that is its sign occurred no more than a decade or two before the new theory's enunciation. The novel theory seems a direct response to crisis. Note also, though this may not be quite so typical, that the problems with respect to which breakdown occurred were all of a type that had long been recognized. Previous practice of normal science had given every reason to consider them solved or all but solved, which helps to explain why the sense of failure, when it came, could be so acute. Failure with a new sort of problem is often disappointing but never surprising. Neither problems nor puzzles yield often to the first attack. Finally, these examples share another characteristic that may help to make the case for the role of crisis impressive: the solution to each of them had been at least partially anticipated during a period when there was no crisis in the corresponding science; and in the absence of crisis those anticipations had been ignored.

The only complete anticipation is also the most famous, that of Copernicus by Aristarchus in the third century B.C. It is often said that if Greek science had been less deductive and less ridden by dogma, heliocentric astronomy might have begun its development eighteen centuries earlier than it did.[15] But that is to ignore all historical context. When Aristarchus' suggestion was made, the vastly more reasonable geocentric system had no needs that a heliocentric system might even conceivably Have fulfilled. The whole development of Ptolemaic astronomy, both its triumphs and its breakdown, falls in the centuries after Aristarchus' proposal. Besides, there were no obvious reasons for taking Aristarchus seriously. Even Copernicus' more elaborate proposal was neither simpler nor more accurate than Ptolemy's system. Available observational tests, as we shall see more clearly below, provided no basis for a choice between them. Under those circumstances, one of the factors that led astronomers to Copernicus (and one that could not have led them to Aristar-chus) was the recognized crisis that had been responsible for innovation in the first place. Ptolemaic astronomy had failed to solve its problems; the time had come to give a competitor a chance. Our other two examples provide no similarly full anticipations. But surely one reason why the theories of combustion by absorption from the atmosphere—theories developed in the seventeenth century by Rey, Hooke, and

---

[14] Whittaker, *op. cit.,* I, 386–410; and II (London, 1953), 27–40.

[15] For Aristarchus' work, see T. L. Heath, *Aristarchus of Somas: The Ancient Copernicus* (Oxford, 1913), Part II. For an extreme statement of the traditional position about the neglect of Aristarchus' achievement, see Arthur Koestler, *The Sleepwalkers: A History of Man's Changing Vision of the Universe* (London, 1959), p. 50.

Mayow—failed to get a sufficient hearing was that they made no contact with a recognized trouble spot in normal scientific practice.[16] And the long neglect by eighteenth-and nineteenth-century scientists of Newton's relativistic critics must largely have been due to a similar failure in confrontation.

Philosophers of science have repeatedly demonstrated that more than one theoretical construction can always be placed upon a given collection of data. History of science indicates that, particularly in the early developmental stages of a new paradigm, it is not even very difficult to invent such alternates. But that invention of alternates is just what scientists seldom undertake except during the pre-paradigm stage of their science's development and at very special occasions during its subsequent evolution. So long as the tools a paradigm supplies continue to prove capable of solving the problems it defines, science moves fastest and penetrates most deeply through confident employment of those tools. The reason is clear. As in manufacture so in science—retooling is an extravagance to be reserved for the occasion that demands it. The significance of crises is the indication they provide that an occasion for retooling has arrived.

[16] Partington, *op. ext.,* pp. 78–85.

Scientific Revolutions and Scientific Paradigms

By John Oakes

You have already read a chapter by Thomas Kuhn from his groundbreaking book, *The Structure of Scientific Revolutions*. In this book, Kuhn shows from the history of science that the course of development of knowledge within a scientific discipline does not progress smoothly, as one might expect from a simple understanding of the scientific method. Kuhn proposes that scientists do not investigate nature by following a simple "scientific method." Instead, he suggests that the course of scientific discovery is governed by a working scientific paradigm, not by application of the scientific method to nature. What he discovered is that in the course of development within a scientific discipline such as chemistry, geology, or biology, one can find evidence for what he

calls scientific revolutions. Such revolutions bring about a very rapid growth of effective knowledge in a discipline, followed by many decades, or even by hundreds of years, of slow and gradual development. When a scientific revolution occurs, there is a complete overthrow of the former scientific paradigm, with a subsequent massive burst of new discovery. New vocabulary is developed, new kinds of questions are asked, and completely new answers are given. Such scientific revolutions are followed by long periods of relatively slow increase of knowledge, which Kuhn calls "normal" science. In this chapter, we will explore this idea using real examples of scientific paradigms and scientific revolutions. First, let us consider some definitions.

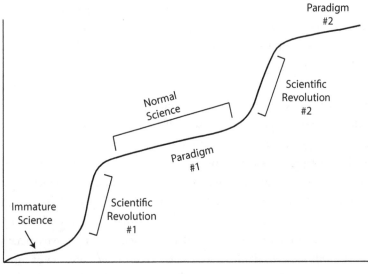

Definition: A *scientific paradigm* is an underlying model or assumption which determines how new discoveries are interpreted and what kind of questions are asked within a scientific discipline.

Definition. A *scientific revolution* is a radical change within a scientific discipline, which occurs when there is a change in the accepted paradigm. The revolution results in a quantum leap in knowledge within the discipline, as well as leading to the advent of new methodologies and vocabulary.

Definition. *Normal science* is scientific research done within the parameters defined by a working model—a scientific paradigm.

According to Thomas Kuhn, most scientific research is not revolutionary and amounts to what he refers to as "problem solving." In other words, most research involves applying the accepted model to an ever-wider array of problems.

Definition: An *immature science* is a scientific discipline which has no working model. It has no accepted paradigm.

Kuhn's study of the history of science led him to conclude that without a paradigm, a scientific discipline such as geology or chemistry or cosmology can make very little progress. An immature science is marked by competing explanations, none of which is yet sufficiently successful to be accepted by those working in a particular discipline of science.

These concepts are best understood by considering real examples. We will look at all the major scientific disciplines and look at the history of knowledge within these disciplines in order to see how Kuhn's description can be used to explain that history. First,

let us consider a list of the current working paradigms in the different scientific disciplines.

Biology:
   Evolution
   The Central Dogma: DNA $\leftrightarrow$ Proteins

Chemistry:
   The Nuclear Atom
   Quantum Mechanics

Physics:
   Quantum Mechanics
   Relativity
   The Unified Model

Cosmology:
   Heliocentrism
   The Big Bang

Geology:
   Uniformitarianism
   Plate Tectonics

Medicine:
   Germ Theory
   Genetics

If Kuhn is right, then these paradigms did not arise out of thin air. Each was produced by a scientific revolution. In some cases, the current paradigm is the original one, while in others an earlier one was the original. An immature science is transformed into a mature one, which, by definition, is governed by a working model/paradigm. In other cases, the paradigm replaces a formerly accepted one. If that is the case, then we have what we will call a paradigm shift. A major portion of this chapter will describe in some detail the story of such paradigm shifts.

As a rule, Kuhn believes that a paradigm shift is a more volatile process than the original scientific revolution in a discipline. The reason is that all the scientists working in a particular discipline have more or less accepted the current model. Their thinking about their experiments, the questions they ask, the types of answers they propose, and the vocabulary they use is all governed by their paradigm. To ask scientists to change their paradigm is truly a great challenge. One of the players in the paradigm shift from classical mechanics to quantum mechanics is Max Planck. He sarcastically said the following about those who clung tenaciously to classical mechanics, despite the overwhelming evidence that quantum mechanics was a superior explanation of physics on the microscopic scale:

> A new scientific truth does not triumph by convincing its opponents and making them see the light, but rather because its opponents eventually die, and a new generation grows up that is familiar with it.

What is it that finally causes the scientific community to embrace the new model, despite their comfort with the former model? Kuhn argues that it was the emergence of anomalies—phenomena which the current accepted paradigm could not explain. At first, those working in a particular discipline either ignore such anomalous discoveries, or they make excuses and patch up their models in ways that will seem to us to be *ad hoc* (i.e., created, not because the evidence leads that way, but in order to defend a particular view). Over time, such anomalies mount so that, finally, a revolutionary scientist proposes a new model—a new paradigm which ultimately, after many years, wins out over the old paradigm.

Consider the following outline of "requirements for a paradigm shift."

1. Anomalies arise; evidence that the current paradigm cannot explain.

2. A young scientist on the fringe comes on the scene.
3. The revolutionary scientist must be a theoretician.
4. The revolutionary makes a shot in the dark.
5. The revolutionary must have character traits such as courage, patience, and communication skills.
6. Dramatic predicted consequences from the new model prove true.

The essential cause of a scientific revolution is that the working model simply cannot explain well-known experimental results. We will see many examples of this. Such anomalous discoveries alone generally are not sufficient to cause a revolution. A scientist on the fringe who is not part of the "establishment" is capable of taking an unbiased look at the problem and propose radical new solutions. We will see one exception, but in almost every case the one to cause a scientific revolution is quite young—generally in their twenties, or at most, their thirties.

One would think that an experimentalist would be the one to create a scientific paradigm, but this is not the case. The evidence requiring a new explanation has already been around for a number of years. New data is not needed. What is required is a theoretician—a person who can create a new explanation. We will see one exception to this rule, but most revolutionaries in science are theoreticians. Does this mean that the theoretician is the most vital member of the scientific community? Perhaps.

The key contribution of the scientific revolutionary is to propose a brand-new explanation. We will see that in every case, the new explanation is truly revolutionary. According to the accepted paradigm, it is not "logical." It is a shot in the dark. If we accept the new model, then nearly everything we thought was true is now no longer true. We will have to stop using much of our accepted vocabulary and start using new words to describe phenomena.

Being a revolutionary does not just require that one have great ideas. It requires that the scientist be willing to take on unstinting criticism. He or she must have very thick skin and be willing to push an idea, despite persecution. Copernicus risked being fired. Pasteur was challenged to a duel. Darwin was vilified by many. To use an analogy, not just any baseball player could have broken the color barrier. It took a special person like Jackie Robinson to successfully bring about a "paradigm shift" in baseball. We will see a couple of scientific revolutions where the one who proposed the idea did not have the requisite qualities, which delayed the scientific revolution for more than one generation.

Another pattern which can be seen in every scientific revolution is that the one who proposes the new model is able to make a prediction using the new paradigm that would have seemed ridiculous or absurd using the old paradigm. When such "dramatic predicted consequences" are observed in the laboratory (usually not by the revolutionary, who is a theoretician!), then finally, the scientific community can be won over to the new explanation.

Consider the list of scientific revolution/paradigm shifts below. We will use the rest of the chapter to explain some of these revolutions using the six-part outline above.

1. Geocentrism → Heliocentrism
2. Two sets of laws of nature (Earth and heavens) → One set of universal laws of nature
3. Young Earth/catastrophism → Old Earth/ uniformitarianism
4. Alchemy → Atomic theory
5. Fixity of species → Evolution of species
6. Immature science (no paradigm) → Germ theory
7. Atomic theory → Subatomic particles
8. Classical mechanics → Quantum mechanics
9. Classical mechanics → Relativity
10. Static Earth → Continental drift theory
11. Uniformitarianism → Catastrophism (a mixed paradigm)

Geocentrism to Heliocentrism

Since ancient times, astronomers and cosmologists were in virtually universal agreement regarding the "heavens." All were in agreement that the Earth is the center of the universe, with the sun, moon, planets, and stars all circling around the Earth. The Greek natural philosophers Aristotle and, most famously, Ptolemy, represented this idea.

There was a tiny minority in the Greek philosophers who actually rejected geocentrism, led by Anaxagoras and Aristarchus. They proposed heliocentrism. This was truly a different worldview from geocentrism, as it placed the sun at the center—not of the universe, but of the "solar" system—with the Earth and other planets circling around the sun. This cosmology is known as heliocentrism, because *helios* is Greek for sun. In Anaxagoras's cosmology, the Earth spins and the stars are fixed in position, which explains the rising and setting of all heavenly objects, including the stars.

Unfortunately, despite the strength of their arguments, the idea of Anaxagoras and Aristarchus was rejected by the Greeks, and it completely disappeared from consideration for 15 centuries. For the sake of simplicity, we will describe the paradigm shift from geo- to heliocentrism, ignoring the work of Anaxagoras and Aristarchus.

The first of our six-part outline is evidence the old paradigm cannot explain. Geocentrism can explain the rising and setting of objects in the heavens. It can also explain solar eclipses fairly easily, if we assume that the moon is closer to the Earth than the sun. However, there were a few observations known to the ancients which could not be explained by the geocentric model. First, there is the fact that the planets vary greatly in their brightness. If the planets are unchanging, self-luminous orbs (which was part of the geocentric model), then how can one explain the fact that Venus and Mars vary greatly in brightness? Then there was the question of lunar eclipses. Geocentrism can handle solar eclipses easily, but if heavenly objects are self-luminous orbs, then how can geocentrism explain lunar eclipses? Add to this a discovery by Ptolemy of something known as retrograde motion. If we plot the motion of Mars as it moves against the fixed stars, it moves from east to west steadily for two or three years, but for a brief couple of months it appears to back-track to the east before continuing its east-to-west motion. Such data seems to fly in the face of the simple geocentric model. Lastly, there is the question of the seasons.

If our outline of requirements for a paradigm shift is to work, then we need a young scientist of the fringe of the discipline of astronomy. This need is supplied by Copernicus (1473–1543). The Polish Nicolaus Copernicus was a student of canon law and medicine who also dabbled in astronomy. He was an outsider in that he was, arguably, the first modern scientist. Trained as a mathematician, he was more of a theoretician than an experimentalist, although he did do some of his own astronomical observations. His "shot in the dark" was to assume, contrary to simple observation, that, rather than the heavenly objects revolving around the Earth, the Earth spins on an axis of rotation. This was truly a shot in the dark, as it does not feel to us that the Earth spins. His heliocentrism could easily explain the changing brightness of the planets, lunar eclipses, and even retrograde motion. Copernicus also provided a superior explanation of the seasons by proposing that the Earth spins on a tilted axis.

In 1517, Copernicus described his view in a little pamphlet he wrote, but he only showed it to a small circle of friends. He appears to have been hesitant to face the likely persecution—and even loss of his job—as a canon of the Catholic Church. He did not publish his ideas openly until 1542, the year he died. There is some debate over whether he lived to see the first copy of his book. Point number five in our outline is that the one to bring about a scientific revolution must have the character traits required to face the opposition and carry the day. It seems that Copernicus did not have the requisite characteristics. Although his book, *On the Revolution of Celestial Orbs*, was published and, almost miraculously, not banned by the Roman Curia, his model languished for nearly a hundred years.

The last point in our outline of requirements for a paradigm change is that the new paradigm generates a predicted consequence, which, for the former paradigm, would be nonsense. If Copernicus was right, then he could predict that other planets, not just the Earth, would rotate on their axes. In addition, heliocentrism, with the planets lit by a central sun, also predicted that planets, like the moon, would have phases.

The revolutionary shift from geo- to heliocentrism was completed by Galileo Galilei (with help from Johannes Kepler). Galileo had the courage and stubbornness that Copernicus apparently lacked. In fact, Galileo was convicted and sentenced to house arrest for the final ten years of his life for publishing his *Dialogue of the Two Chief World Systems* in 1632. His earlier *Starry Messenger* included observations which confirmed in dramatic fashion the predictions made by Copernicus one hundred years before. Using the recently invented telescope, Galileo observed phases of Venus and the rotation of the sun and moons of Jupiter (something even Copernicus would not have been so bold as to predict).

Separate Laws for the Heavens and the Earth to Universal Laws of Nature

It is arguable that physics itself was an immature science before the career of Isaac Newton. The ancients, especially the Greeks, saw the heavens as subject to different laws than those governing processes on the Earth. To Plato and Aristotle, the heavenly objects were perfect and unchanging, whereas things on the Earth changed and decayed. As late as the early 17th century, Johannes Kepler proposed that angels moved the planets through the heavens in elliptical orbits. The evidence the former paradigm could not explain in this case is the motion of heavenly objects. Planetary motions could be described, but not explained.

Isaac Newton (1642–1727) fits the ideal description of a scientific revolutionary. He was trained as a mathematician, not a scientist, making him an outsider to the scientific community. He was first and foremost a theoretician, as required by our outline. He published one important experimental discovery about the action of prisms to separate light, but the rest of his scientific publications were mathematical and theoretical. As is consistent with those sparking scientific revolutions, most of Newton's major discoveries were at least initiated by 1666, when he was 24 years old.

Newton's long shot was a radical proposal. It involved his famous explanation of the falling of an apple. The common understanding is that Newton "discovered" gravity. This is not exactly correct. Newton's discovery was that the force which draws the apple to the Earth is the same which causes the moon to circle around the Earth. Newton proposed the UNIVERSAL law of gravity, and, by implication, he proposed that all the laws of nature are universal. There are not two separate sets of laws for the heavens and the Earth, and experiments done on the Earth can be used to model events in the farthest heavens. Newton, in effect, discovered the idea of the Mechanical Universe. The entire universe runs according to natural laws that do not require the scientist to invoke supernatural explanations. Newton demystified the universe. Some have described Newton's discovery of the universal laws of nature and the Mechanical Universe as the beginning of the modern era.

Newton had more than sufficient stubbornness and communication skills to bring to fruition this scientific revolution in his own lifetime. Although not very good at social skills, Newton's book, *Principia*, published in 1687, was so brilliantly composed that it ultimately converted even his critics to his belief in universal laws.

No scientific revolution is complete without a dramatic predicted consequence that proves true. In the case of Newton's universal law of gravity, it was his friend (of whom Newton had few!) Edmund Halley who took on this role. Using Newton's equations of motion, he was able to predict an elliptical orbit for what is now known as Halley's Comet. Halley's calculations in 1705 predicted a period of orbit of 75 years. He was also able to show that, sure enough, a comet had been observed every 75 years in the past—a dramatic prediction which explained one of the greatest mysteries of the ancients. Comets were no longer random and unexplained heavenly events, but objects whose motion could be explained by universal laws of motion.

Young Earth/Catastrophism to Old Earth/ Uniformitarianism

Our description of the third scientific revolution concerns the accepted model of the age and history of the Earth. It involves the creation of one of the two principle paradigms of modern geology—uniformitarianism. The other main paradigm of geology is plate tectonics. The creation of this paradigm is described in the section of the text on pseudoscience.

As with the creation of the universal law paradigm, it is debatable whether the former young Earth paradigm was a scientific one, and that, therefore, uniformitarianism was preceded by an immature science. We are calling the geological model which preceded the work of James Hutton young Earth/ catastrophism. It is doubtful that anyone in the 18th century would have used this label. However, if we were able to ask the majority of western Europeans in the 18th century, including scientists, how old the Earth is, most would respond that it was only several thousands of years old. The reason this is a questionable scientific paradigm is that this belief was based primarily on a religious presupposition, rather than on any scientific evidence for such a young age of the Earth.

The age of the Earth is one thing, and its history is another. In the 18th century, proto-geologists generally accepted what we now call catastrophism. In other words, they believed that the features of the Earth were created more or less as they are today, but that it was also affected by massive global, world-changing catastrophic events, to which there is no analogy happening within human memory. The most famous of these catastrophes, of course, was the universal flood described in the biblical book of Genesis. However, most who published in the field of what we would call geology today also believed in other world-changing events in the somewhat distant past.

There were two pieces of evidence that emerged in the 18th century which ultimately resulted in the overthrow of the former worldview. These were sedimentary rocks and fossils. Of course, both had been observed long before the 18th century, but by the 1700s, amateur geologists began to collect and categorize both different kinds of rocks and a variety of fossils. That these both imply great length to the history of the Earth was bound to emerge eventually.

The scientist on the fringe who created what we are calling old Earth/uniformitarianism is James Hutton (1726–1797). For good reason, he is called the "father of geology." He invented the names for three principle categories of rocks—sedimentary, igneous,

and metamorphic. Despite the fact that he deserves this designation, Hutton was a true outsider (point #2 in our outline). He had relatively little training as a scientist, although he did earn a medical degree. In fact, his principle occupation was as a farmer. He was in a circle of intellectuals in Edinburgh, who created what is sometimes called the Scottish Enlightenment. Among his colleagues in Edinburgh were David Hume and Adam Smith. Was he a theoretician (#3 in our outline)? Hutton did much observation of the rock formations around the Scottish countryside, but did no experiments. His chief skill was observation and theorizing.

Hutton's gamble was to propose that the Earth is very old. How old? To quote the father of geology, "no vestige of a beginning, no concept of an end." Hutton had the audacity—despite the accepted religious presupposition—to propose that the Earth is so old that there is not even a vestige of the original geologic formations. Hutton saw the millions of sedimentary layers, many of which have been deformed at bizarre angles from their original horizontal deposition, as evidence of what is called "deep time." He proposed that the cause of geologic change is heat escaping from within an initially hot Earth. Hutton principally argued from sedimentary rocks, but he also considered fossils ("figured stones" to Hutton) to be strong evidence of change over vast periods of time.

The most important concept invented by Hutton is uniformitarianism.

The definition of *uniformitarianism* is the belief that the physical features of the Earth were created by slow, gradual processes which are observable today, acting over vast ages of time.

Unfortunately for the development of geology, Hutton lacked some of the traits needed to bring about a scientific revolution. His friend, John Playfair, claimed that Hutton had these ideas in mind as early as 1760 while in his thirties, but Hutton was slow to publish. He did little to publicly defend his controversial uniformitarianism. Hutton published his ideas in the book, *Theory of Earth*, in 1788. The problem with this seminal work is that Hutton's writing style is impenetrable. A Hutton sentence can last for an entire page and a paragraph for two or three. When he died, only a small circle in Scotland and a few like-minded in Germany had been influenced by uniformitarian ideas.

Hutton's dramatic predicted consequence from his theory was that entire regions of the Earth's surface rise and sink at an infinitesimal rate, which, extrapolated over millions of years, could raise ocean-laid sediment to the top of the highest mountains in Scotland. The rise of land out of the sea was first demonstrated in the 19th century when it was proved that Scandinavia is rising out of the Baltic and North seas.

Although Hutton did not have the qualities required to convince most geologists of old Earth/uniformitarianism, his ideas were eventually accepted by all scientists. Actors in this drama include Hutton's close associate, John Playfair (1748–1819), and English civil engineer William Smith (1769–1839). Smith invented the idea of index fossils. Fossils found deeper in the fossil record are relatively older. As Smith said, each layer had "fossils peculiar to itself." Most important to the acceptance of uniformitarianism was the work of Scottish geologist Charles Lyell (1797–1875). Lyell was the most influential geologist in the 19th century. His *Principles of Geology* (1829) became the standard text of geologists throughout Europe and North America, despite the fact that many of his readers believed in the young Earth perspective. Lyell famously said, "The present is the key to the past." By 1850, the battle over the age of the Earth was all but complete, and the scientific revolution begun by Hutton was complete. We will see that Lyell's unformitarianism influenced Charles Darwin in constructing his theory of evolution.

Fixed Species to Evolution of Species by

Natural Selection

The fourth scientific revolution we will use to demonstrate Kuhn's description of the history of science involves the origin of species. The working assumption of scientists in the 18th and the first half of the 19th centuries is known as the fixed species model. According to this view, living species change only very little over time. This model could allow for a species to become extinct, but if correct, there should be no new species created over time, and the fossil record should reflect this proposition. The most well-known botanist and chief supporter of the fixed species paradigm in the 18th century was the Swedish botanist Carl Linnaeus (1707–1778). Linnaeus believed in miraculous individual creation of each species by God. This was the accepted presupposition of botanists well into the 19th century.

The "evidence the old paradigm could not explain" in this case is fossils. Up to the time of William Smith, it was principally amateurs who collected fossils, called "figured stones" by James Hutton. Only gradually did it become clear that the fixed species concept would not explain the fossil evidence, at least in part because paleontology was not a well-established scientific discipline at that time. When the odd fossil was discovered, naturalists would propose that it was a larger or smaller version of a present species or that it came from a not-yet-discovered species. As late as the mid-19th century, large portions of the Earth had not yet been explored by Europeans, making this a more reasonable explanation than it would be today.

The young scientist on the fringe of biology who created the evolutionary paradigm, of course, is Charles Darwin (1809–1882). His was not the first theory of evolution. That distinction was held by Jean Lamarck (1744–1829). The French naturalist published what he called vitalism theory, or the theory of acquired characteristics, in 1801. According to Lamarck, species have an innate ability to strengthen whatever traits are used consistently over time. For example, Lamarck believed giraffes gained their long necks by continually stretching them, generation after generation. Unfortunately for this theory, it was never supported by experiment, so it eventually was rejected.

Darwin qualifies as a scientist on the fringe of botany, given that his degree from Cambridge University was in preparation for the Anglican priesthood. His interest was in natural theology. He studied under the preeminent naturalists of the day: John Henslow and Adam Sedgwick. They encouraged him to take his natural direction, which was natural history. Henslow talked Darwin's father into letting him join a voyage of discovery on the HMS *Beagle* in 1831, when Charles was just 22 years old. It was on his famous five-year journey on the *Beagle* that Darwin blossomed into the preeminent naturalist of his age. Over five years exploring South America and the South Pacific, Darwin sent back thousands of plants and animals and geological (and even archaeological) samples. He took with him a copy of Lyell's *Principles of Geology*, which influenced him as he considered his discoveries concerning living species. Darwin discerned slow change of species in parallel to Lyell's slow change in geology. One of Darwin's discoveries was relevant to the work of Hutton, as he noted that the continent of South America was very slowly rising out of the Atlantic and Pacific oceans.

It was on the Galapagos Islands that Darwin made his most famous observations. In order to simplify a vast array of discoveries which led to Darwin's greatest proposal, we will use his observations of the Galapagos finches (his work with mockingbirds, tortoises, and iguanas was also key). Being the brilliant and systematic naturalist that Darwin evolved into, he noted 14 different species of finch on the islands. The different species were uniquely adapted to the available food in the wide variety of microecosystems on Galapagos Islands. There was a toucan-like finch that ate fruit and another species with sharply pointed beaks for eating nuts, as well as ones suited for eating the small seeds in the savannas. Darwin returned to England after a five-year voyage. He claims to have only been convinced of natural selection and the survival of the fittest after he returned to Cambridge.

To simplify a diverse range of evidence Darwin eventually brought to the question of the origin of species, let us describe his shot in the dark in the following way. Darwin concluded that a single species of common South American finch with a wide variety of available traits arrived at the Galapagos. Finding an environment with an abundance of food, the finches

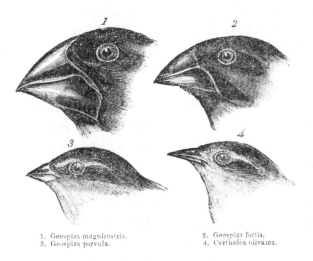

1. Geospiza magnirostris.
3. Geospiza parvula.

2. Geospiza fortis.
4. Certhidea olivacea.

Darwin's finches.

Copyright in the Public Domain.

experienced a population explosion. When available food became scarce, natural selection favored different traits in different ecological niches. Over thousands of generations, natural selection among the wide variety of available traits produced the 14 separate species of finch observable on the Galapagos today. Darwin enlarged this example and proposed this to be the explanation of how all species on Earth were produced by a gradual process over many millions—or even billions—of years.

Like Hutton, Darwin was a hesitant theoretician. He sat on his principle discovery for more than 20 years. In the meantime, he included studies of unnatural selection in the selective breeding by farmers in his thinking about the change of species over time. It was only when a young upstart naturalist named Alfred Russel Wallace prepared to publish a theory similar to Darwin's speculations that he was finally convinced to publish his theory. Before publishing his great work, the gentleman Darwin published a short paper with Wallace on natural selection. Just months later, Darwin published his famous book, *On the Origin of Species*, in November 1859. It was an immediate sensation. In the introduction, Darwin summarizes his theory as follows: "As many more individuals of each species are born than can possibly survive; and as, consequently, there is a frequently recurring struggle for existence, it follows that any being, if it vary however slightly in any manner profitable to itself, under the complex and sometimes varying conditions of life, will have a better chance of surviving, and thus be *naturally selected*. From the strong principle of inheritance, any selected variety will tend to propagate its new and modified form." The rest, as they say, is history.

Did Darwin have the courage, stubbornness and communication skills needed to carry out a scientific revolution in his own lifetime? The simple answer is yes. Although hesitant to take on his harsh critics, and despite being unwilling to speculate on human evolution because of the potentially explosive nature of the

A note from Darwin's notebook in 1837, in which he speculated over the common descent of species.

Copyright in the Public Domain.

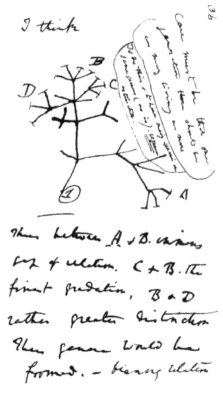

topic, Darwin published a carefully reasoned, brilliantly argued book with sufficient experimental support that, by its own power, carried the day. Darwin's "bulldog," Thomas Huxley, took on the public debate that the sickly Darwin could not withstand.

A successful scientific revolution must have a dramatic predicted consequence if our working outline is correct. Darwin proposed in his book that, over time, the fossil record would be found to contain innumerable transitional fossils—that the "missing links" would no longer be missing. The process of filling in the fossil record has been slow and painstaking, but on the whole, Darwin's many predicted consequences have proved true. Very helpful, of course, was Gregor Mendel's discovery of genetics as a means of passing information from parent to offspring.

Immature Science/No Paradigm to Germ Theory

The mid-19th century was a fruitful time in the history of science. It saw the origin of the two chief paradigms of biology: evolution and genetics. It also produced the first paradigm of medicine. The next scientific revolution we will study moved medicine from an immature science to the creation of its first successful paradigm.

The "evidence which could not be explained" in this case is infectious disease. Throughout human history, the chief cause of death had been the outbreak of infectious diseases. The search for the cause of disease had been a major pursuit of humans from ancient times. Without the scientific method, the cause of disease was bound to escape humans indefinitely. Naturally, the cure for such diseases also eluded humanity. Kuhn describes the period when a discipline was an "immature science" as a time of competing models. You have already read the wonderful article about John Snow's discovery of the cause and cure of cholera. In this article, you saw this chaotic interaction between competing explanations in play. This supports Kuhn's ideas. Some proposed that sin was the cause of disease. Others theorized "effluvia," or bad air, to be the cause of epidemic death. Still others invoked the presence of certain minerals in the diet or the alignment of the planets.

The "young scientist on the fringe of medical science" in this case was Louis Pasteur (1822–1895). Others such as Joseph Lister, John Snow, and Ignaz Semmelweis played a role as well, but Pasteur was the chief antagonist in the battle over the cause of disease. The French scientist was trained, not as a biologist or medical scientist, but as a chemist. One can argue that if Pasteur had been trained in traditional medical science, he might never have proposed germ theory. Pasteur is an exception to our rule that the one to bring about a scientific revolution will be a theoretician. Pasteur is one of the greatest experimentalists in the entire history of science. His work virtually created the science of microbiology. His discoveries in chemistry are massive as well.

Pasteur's work led him only indirectly to his greatest discovery—to his shot in the dark—that disease is caused by microbes, which invade the bodies of animals and can produce illness and even death. His germ theory was preceded by his demonstration that yeast microbes are the cause of fermentation. He also proved that it was bacteria, which had for millennia been the cause of the spoiling of a portion of fermented grape juice. His discovery of the process that we now call pasteurization to kill bacteria was first applied to saving the French wine industry (making Pasteur a hero of France!).

By the 1860s, Pasteur realized that bacteria and other microbes were the cause of animal disease. He isolated the bacteria that cause animal anthrax and created a successful immunization for anthrax. Scientific revolutionaries always experience vigorous resistance from the mainstream scientists within the discipline. This pattern is well illustrated by the reaction to Pasteur's germ theory. Pasteur was not unwilling to confront his opponents. He was very self-assured and combative to his opponents. One offended surgeon even challenged him to a duel. Pasteur had all the character traits required to bring about a scientific revolution, including a dogged determination and a famous stubbornness in defending his discoveries.

Pasteur gave us a wonderful example of a "predicted consequence which proved true." Despite years of research, he failed to discover the microbe that causes rabies, that devastating disease. We now know the reason he was unable to discover a living microbe as the cause of rabies; it is because rabies is a viral disease. Viruses were too small to be observed by any microscope before the 1940s, when electron microscopes were developed. Pasteur proposed the existence of a microbe that could not be seen—thus indirectly discovering what we now call viruses. He applied his germ paradigm to this unexplained infectious disease. Working with his colleague, Emile Roux, Pasteur dehydrated the spines of rabid rabbits, producing a vaccination which prevented those bitten by rabid animals from dying of rabies. His first successful patient was nine-year-old Joseph Meister. Not only did the boy survive, he worked for the Pasteur Institute until his death in 1940—committing suicide rather than let the Nazis desecrate the grave of his savior, Pasteur.

Classical Mechanics to Relativity

By the mid-19th century, scientists such as Lord Kelvin (William Thompson) began to declare that most of the essential problems in physics had already been solved. We had reached a relatively boring time of merely applying the great truths to ever-more-complex versions of the same kinds of problems. The career of the most famous scientist in history proved this prediction to be a massive error. We are talking about the man who had a bad hair day every day—Albert Einstein.

First, let us describe classical mechanics as it relates to the paradigm that replaced it—relativity. Classical mechanics was, essentially, the physics of the 19th century. It included a number of laws and theories about motion, particles, and waves. Relevant for this study is what classical mechanics predicted about measurements, which should be absolute, versus measurements which should be relative. A measurement is "absolute" if it is independent of motion. A measurement that is "relative" is one whose value will change when motion is occurring. For example, classically, we expect that the mass and size of an object should not be affected by whether the object is moving. We would say, then, that mass and length or distance are absolute under classical mechanics. Similarly, we expect, classically, that the rate at which time passes should not change if we are moving. Time, then, is absolute, at least according to classical mechanics. In

fact, one could argue that it is "obvious" that mass, time, and length are absolute.

Not all measurements, however, should be absolute, according to classical mechanics. For example, imagine you are in a train traveling at 30 miles per hour. Imagine also that a car is passing the train going at 50 miles per hour. What speed would you, on the train, measure the car to be traveling at? The answer, of course, is that you would measure only 20 miles per hour. Why? Because the measurement of velocity is relative. In fact, if the car were approaching the train rather than passing the train, you would measure 80 miles per hour. As another example, consider a motionless object on the Earth. Its speed is zero. Correct? No, actually it is traveling as much as 1000 miles per hour, depending on how close to the equator it is located. Clearly, the speed of an object is indeed "relative."

What about our measurement of the speed of light? Classically, it too must be relative. If light is traveling toward you at the speed of light, but you are also moving toward the source, "obviously" you should measure a speed slightly higher, by analogy to the car/train example above. Similarly, if you are moving in the same direction as the light, you should measure a slightly lower speed of light.

Here is what we expect from classical mechanics:

Things that are absolute	Things that are relative to motion
mass, time, length/distance	speed of an object, speed of light

An important aspect of classical mechanics in the 19th century was known as ether (or aether) theory. It was believed at the time that all waves must pass through some sort of medium. Light is a wave; therefore, it must pass through something. The thing through which light travels is, by definition, ether. If the ether theory were true, then, although velocity measurement would still be relative to motion, there would be a means to define the absolute velocity of an object, which would be the velocity of that object in the nonmoving ether. If the ether theory were correct, then the speed of light would not depend on the motion of the source of light, but it would depend on the motion through the ether of the one observing the light.

This brings us to the anomalous "evidence the old paradigm could not explain." In 1887, a pair of physicists, Albert Michelson and Edward Morley, measured the speed of light of two beams moving perpendicular to one another. If the ether theory were correct, because the Earth is moving through the ether, they should have measured very slightly different speeds of light for the two beams. In fact, the result was a negative experiment, much to the shock of Michelson and Morley—and of all scientists at the time. The speed of light was the same, no matter the direction. The ether theory was thus falsified. Let us put the result of the Michelson/Morley experiment simply. *The speed of light is independent of motion of source or observer.* In other words, the speed of light is absolute. The result was to move the speed of light from the relative to the absolute column in the table above.

What was the scientific community to do with such a clearly anomalous result? The answer is that they struggled with this experimental finding for almost 20 years, until a young scientist on the fringe named Albert Einstein (1879–1955) proposed a solution. Einstein was "on the fringe" in that he only had a bachelor's degree, with training as a mathematics teacher. Upon graduating, he was unable to find a teaching position, at least in part because he had not achieved top grades at Zurich Polytechnic University. He ended up working in a patent office where, in 1905, at the tender age of 26, the young scientist published four monumental papers in a single year. One of them launched the scientific revolution that produced the paradigm known as quantum mechanics. But that is another story. Fitting our "requirements for a paradigm shift," Einstein was the ultimate theoretician in that he did no experiments at all.

In his famous paper on special relativity in 1905, Einstein made one of the most dramatic shots in the dark in the history of science. He proposed that physicists should suspend nearly everything they knew from classical mechanics. He took as his only postulate the Michelson/Morley conclusion that the speed of light is absolute. With this assumption and using F = ma (Newton's second law of motion), Einstein made three striking predictions. First, he proposed that the amount of time which one experiences depends on how fast one is moving. The faster an object moves, the less time it experiences. This stretching of time is called *time dilation*. Einstein derived a second prediction from his postulate that the speed of light is absolute. His second predicted consequence was that our measurement of distance depends on how fast we are moving. The faster an object moves, the smaller it appears to be. This is strikingly counterintuitive. Most shocking of all, Einstein derived from his postulate that mass itself is relative. His conclusion was that the faster an object moves, the more mass it acquires.

Our table of what is absolute and what is relative is now quite different.

	Measurements that are absolute	Measurements that are relative to motion
According to Classical Mechanics	mass, time, length/distance	speed of object, speed of light
According to Einstein's special relativity	speed of light	speed of object, mass, time, length/distance

In November of 1905, Einstein published a second paper of theoretical results from his postulate of special relativity. When studying the implications for conservation of mass and energy, he concluded that, essentially, mass is a form of stored energy. This resulted in the most famous equation in all of science: $E = mc^2$. Ultimately, this prediction led to the production of massive energy from atomic fusion and fission.

Did Einstein have sufficient patience, courage of his convictions and communication skills to bring to completion this scientific revolution? The simple answer is yes. This happened over the next ten years or so, despite the fact that time dilation, length contraction, and mass increase had not yet been directly demonstrated in the laboratory. Time dilation eluded measurement in the early 20th century because until one comes quite close to the speed of light, the size of the dilation is way too small to measure. Albert pushed on. In 1916, he published his broader general theory of relativity. This paper is more general, in that it allowed for acceleration of an object (a possibility not considered in his 1905 paper). The mathematics of the general theory of relativity is rather difficult, which explains the intervening 11 years. Essentially, Einstein's general theory of relativity is a theory of gravity. Einstein concludes that objects with mass cause a "warping," a curving of space. The immediate practical implication of this speculation is that light itself could be bent by sufficiently massive objects, even though light has no mass.

It turns out that this implication of general (not special) relativity became the dramatic predicted consequence which convinced the entire scientific community of Einstein's new paradigm. In 1919, Arthur Eddington traveled to Principe, an island in the South Pacific, where a total eclipse of the sun that year was predicted. His intention was to test the prediction of Einstein's general theory. If Albert was correct, then a star just behind the sun should be able to be seen immediately next to the sun because of the slight bending of the light from that star as it passed near the sun. Of course, such an observation could only be made during a total eclipse. Einstein's prediction proved true, and even the amount of bending of light was equal to Einstein's prediction. By 1920, Einstein was triumphant. He had become the most famous scientist in the world. The scientific revolution was complete.

Summary

A quick glance at the list of scientific revolutions earlier in this chapter will show that we have not yet covered all the great scientific revolutions. Space constraints will not allow us to discuss what is one of the most essential changes in scientific paradigms from classical to quantum mechanics. The change from static Earth theory to continental drift/plate tectonics will be used as a case study on the distinction between mainstream science and pseudoscience. The recent revival of catastrophism due to discovery of the Chicxulub asteroid in the Yucatan Peninsula and to surprising discoveries surrounding the Cambrian explosion is a fascinating topic as well. Two of the three important revolutions in the history of chemistry have not been covered here. Perhaps these will be discussed in the course you are taking.

However, the examples already mentioned do make the point fairly well. The scientific community works most comfortably within an accepted model/paradigm. Most of science is "normal" science—that is to say, it is problem solving. As a rule, scientists strenuously defend their paradigms, even in view of what seems with hindsight to be rather obvious anomalies.

Revolutionary ideas are initially rejected in every case. The greatest discoveries of science are caused by revolutionaries—the few genius outsiders with the courage and communication skills to drag the scientific community forward, kicking and screaming. The sordid relationship between the establishment and science's revolutionaries, if nothing else, makes for fascinating reading.

Section IV
Ethics

Ethics and Science

By John Oakes

Should a scientist be allowed to pursue research into the cloning of human beings? Even if it is legal to do so, ought an individual human being who happens to be a scientist pursue developing such technology? Should our society as a whole hold those who make new discoveries responsible for the ethical implications of their research? Is it a good idea to let scientists police themselves in these areas, on the assumption that they know the scientific implications better than the layperson? Or would it be wiser to insist that nonscientists have a major role in deciding what kinds of research projects companies, governments, and university research labs should pursue? How do we know that the results published by scientists are real data—that a scientist has not faked the data in a selfish pursuit of credit in the scientific community or for potential profit? Even if legitimate scientific work has been done, how can we assure that an unbiased interpretation of the data has been made? Might biased interpretation of research results be dangerous in some circumstances? Should we trust scientists to police their own communities with regard to plagiarism, faked data, and biased interpretation?

All of these questions involve the interface between science and human ethics. Presumably the reader already has a good working definition of science. How are we to define ethics, and how will we evaluate the role of ethics in making decisions about what research to do, and how such research should be conducted? To put it simply, **ethics** is an attempt to balance the positive and negative effects of human behavior on individuals and on society as a whole. The question of ethics can be applied in a broader sense, to include the positive and negative impact of human actions—not just on other humans, but on the earth as a whole, which includes ecosystems and global systems.

One thing that can help us to understand the realm of ethics is to distinguish it from morality. Many are confused about these terms, thinking they are indistinguishable. This definitely is not the case. Morality has to do with questions of absolute right and wrong. Unlike ethics, morality does not ask questions of cost and benefit, it asks questions of right and wrong/good and evil based on some sort of authority, such as an accepted scripture or religious teacher, from which such moral laws emanate. In this section, we will not consider questions of morality. Questions of this sort are best dealt with in a religious context. All human beings, no matter their philosophy or religion, can have a common discussion on ethics, but we will not find a unanimously agreed-upon morality. In fact, many people do not even believe that absolute moral judgments can be made. So we will put aside moral considerations, not because they are unimportant or not real, but because it is out of place in our present discussion. Instead, we will consider ethical questions.

Two Areas of Ethics

We will divide our discussion of ethics as it relates to science into two broad, but rather easily distinguished, categories. The first will be the ethics of how scientific research is done and scientific knowledge is advanced. I like to call this "internal ethical issues." We might call it the ethics of doing science. In other words, one large area of ethical concern is how one ethically acquires data, treats that data, and reports one's findings. Broad areas of concern here include the following: plagiarism, faked data, and biased interpretation of data. All three are issues of honesty. These have to do not with the ethical implications of the discovery, but the ethics of how such discoveries are made and reported in the first place.

The second broad area we will discuss is the ethical implications of scientific discovery. Many new technologies developed by scientists allow human beings to do things that may be harmful, either to entire populations, to individual persons, or possibly to the broader environment of the earth as a whole. It is possible for research to be done with impeccable ethics, yet for the result to raise major ethical ramifications for both scientists and nonscientists. I like to call these "external ethical issues" of science. These are ethical concerns that apply to scientists and nonscientists alike. They are human issues. Should we pursue the cloning of human beings? What about the ethical implications of genetically modified crops, either for human societies or for ecosystems? What are the pros and cons of these research programs, and on balance, should we restrict such research—or even ban it outright?

Ethics: Internal Issues

We will first discuss the ethics of doing science. What is the most ethical means of doing research, and perhaps more importantly for us nonscientists, how do scientists ensure that unethical behavior is minimized or prevented? The first, and arguably, the *only* ethic of "doing" science is honesty. This has been the case almost from the inception of the scientific revolution. We can go back to the time of the Lunar Society in England, which became the Royal Society when it transitioned from an underground group of scientists meeting in secret, to become the first royally commissioned society of scientists in history. From the beginning, the culture of science included the reporting of one's research findings to a community of fellow scientists for scrutiny. Results became an accepted part of the scientific community only after they were subjected to sometimes withering criticism by one's peers at the meetings of the Royal Society. Those whose works could not be reproduced by others in the society were put to shame and ultimately were removed from the society of fellow scientists. From the beginning, it was clear that unless the work of fellow scientists could be trusted to have been conducted with absolute honesty, scientific progress would be greatly hindered. We can imagine a scientist preparing to present his work to his fellow researchers, searching his data and his conclusion for any signs of weakness. The question of priority—who was to get credit for the discovery—was vigorously debated in such societies. This is how the culture of science learned to enforce ethical behavior in the ranks of those accepted to this relatively closed society of scientists.

Let us separate the ethical issues raised by the scientific process itself into three areas:

1. **Plagiarism**: The use of another's published writing or intellectual work in order to gain credit as if it were one's own work, without acknowledging the source.

2. **Faked Data**: The willful manipulation of data gained by experiment or the willful fabrication of such data, for the creation of the false impression that the data is legitimate.
3. **Experimental Bias**: This occurs when a scientist allows his or her expectations for an experiment to influence the interpretation of the data. It can involve either biased recording of observations or biased interpretation of data.

Why would a scientist commit one of these unethical behaviors? The answer is that there are a variety of causes of such dishonest behavior. First, and perhaps foremost, such things happen because scientists are human beings. Selfishness, greed, and pride are possible motivations for unethical behavior. It is my personal experience that, on average, scientists are probably just a bit more sensitive to guard against unethical behaviors than humans as a whole. However, being just slightly better than their fellow humans on average (assuming that my unscientific assessment is even accurate) leaves a lot of room for dishonesty. We will look at a number of case studies of plagiarism, faked data, and biased interpretation. Some of these will include outright dishonesty by unethical people trying to lie in order to achieve personal gain. However, this is not the whole story.

Publish or Perish

Another cause of unethical behavior is the extreme pressure put on all scientists involved in the game of science in what is commonly known as the "publish or perish" syndrome. In order to do research, financial support is required. In order to obtain tenure at a research university, one must produce as much published research as possible. The pressure on research team leaders to produce a minimum acceptable number of research papers in major journals within a short time span is enormous. If you want to start a one-sided conversation with a scientist, ask him or her to explain the meaning of "publish or perish." Insufficient results, defined as not producing a certain number of publications in peer-reviewed journals within a certain amount of time, will result in a scientist either losing funding or not receiving tenure. Either result is the death knell for the career of that scientist. The pressure to cheat or cut corners is so great that it is hard for a scientist to explain to those outside the club. This pressure, combined with what we might call human nature, produces a wide array of dishonest, unethical behaviors by scientists.

Checks and Balances

As already mentioned, the scientific culture developed the means to largely exclude unethical/dishonest science very early in its history. These do not always work in individual cases, but on the whole, they tend to be relatively effective at preventing dishonest work being published and accepted into the scientific community. We will list a few of these honesty-enforcing measures, which are all a standard part of how scientists police themselves. Scientists will differ somewhat in their evaluation of how effective these measures are, but even the most skeptical will probably agree that without these self-correcting measures, science would have many more ethical problems and would not progress as rapidly as it has. Three principal means of enforcing scientific integrity are as follows:

1. The scientific grant process is peer reviewed and is extremely competitive.
2. Journal publications are peer reviewed.
3. Important results are checked by other scientists for reproducibility.

The latter two are preventive measures, whose role is to keep unethical science from being accepted as legitimate science. The first can act more as the punishment part of the system. The end result is, without research grant money, science in the 21st century simply cannot happen. Research costs money—too much money for virtually any scientist alive today to pursue as a hobby. The days of the gentleman scientists, such as Robert Boyle or Charles Darwin, pursuing their own research as a self-funded hobby, are long gone. The granting of research funds, either from private or public sources is fantastically competitive. Typically, a funding agency might receive ten proposals for each one funded. Probably a large majority of the proposals are good science, but only a small percentage will be funded. Research proposals are received by the granting agency, such as the National Science Foundation (NSF) or the National Institutes of Health (NIH), or a private foundation such as the Petroleum Research Foundation (PRF). The proposals are sent out to scientists who are experts in the particular field for comment.

Here is how the system works: If an individual scientist has been shown to behave unethically, plagiarizing, faking data, or even being quite blatantly biased in the interpretation of data, that scientist will never again receive funding—the end of a career, in effect. The scientific community of peers in a particular discipline is a much smaller club than one might expect. Scientists cannot hide. One and done, that is the rule. This is not "three strikes and you are out." The granting process is the hammer of enforcement of ethics in scientific research.

Then there is the peer review process in science. A scientist cannot gain recognition and cannot influence the course of science as a whole without publishing in journals. For the most important and influential science journals, peer review is part of the process. A proposed research paper is presented to an editor, who sends it out to a small number of peers. These reviewers give a thumbs-up, thumbs-down, or preliminary approval, which requires the author or authors to respond to the criticisms of the reviewer. This particular "check and balance" is particularly good at eliminating biased interpretation of data.

Peers can often detect bias—wishful thinking—that the one doing the research cannot see. This truism applies to bias in all aspects of human activity, but it is particularly useful in science to prevent dishonest, faulty, or unwarranted conclusions being drawn from experimental data.

Last on our list of checks and balances for now is that bogus, dishonest science can be detected by attempts to reproduce the work. The norm in science is that the one doing research to extend knowledge in a particular area will first use their instruments or methodology to reproduce previous important work. After showing that the measurements agree with prior work of others, then the scientist will apply the technique to a new situation. However, as we will see in our case studies, sometimes attempts to reproduce previous work actually ends in showing that the experimental work was done poorly. Some research results are not reproducible! In this case, the one who published the previous work is often required to print a retraction. Either way, this particular "check and balance" is particularly useful at preventing faked data. We are oversimplifying here to some extent. Reproducibility, or lack thereof, can occasionally catch biased interpretation and peer review can detect faked data at times, but the general rule is that peer review prevents biased interpretation, and attempts at reproducing previous results tend to prevent faked data entering the scientific literature.

We will now present a few brief "case studies" of unethical science and how this was detected, but before doing so, a couple of other symptoms or predictors of unethical science bear mentioning.

The first of these is **conflict of interest**. When the one doing research has a financial or other incentive beyond the normal desire to explore the laws of nature, then this can be a red flag for likely unethical behavior. Of course, in all science there is at least some possible financial gain, as the job of the scientist may be on the line, but scientists are generally required to avoid blatant conflicts of interest. Those checking the efficacy and safety of new drugs for a pharmaceutical company should not be paid directly by that company for such experimental work. This would be a clear conflict of interest. Indeed, medical science has not always held to this standard. Ethical lapses have occurred in many cases.

Another red flag for unethical behavior is when a scientist **works in isolation**. Any scientist working alone, analyzing his or her data without colleagues to bounce ideas off of, is particularly prone to wishful thinking. We will see examples in our case studies in which working in isolation has been a contributing factor in unethical behavior.

Plagiarism

Plagiarism in science takes two forms. The first of these involves a scientist using the data or the writing of another scientist and claiming it as his or her own. This is a blatant violation of ethical standards. It is also hard to get away with and has, for this reason, been relatively rare in the history of science. The second kind of plagiarism is more subtle. In such plagiarism, a scientist uses the ideas of another, rather than the actual data or writings of another without giving credit. Such unethical behavior has often occurred within a single research group, including research professors taking credit for the ideas of the graduate student or other researchers working under them. Either way, the motivation for plagiarism is desire for credit, fame, or the perceived need to get an advantage in achieving tenure or capturing grants.

As already stated, blatant plagiarism is rare in the history of science. Generally, even unethical persons are usually not willing to take the chance of stealing the work of another in this way, for the simple reason that the likelihood of getting caught is so high. Two famous examples in the history of science both show how this can happen, and reveal why this kind of behavior typically is unusual.

Our first case study comes from the 18th century. It involves a man named Johann Bernoulli and his son, Daniel Bernoulli. Johann, the father, was a leading figure in Europe in mathematical physics. Apparently Daniel, his son, was even more brilliant than he was. His father insisted that Daniel not become a mathematician. Daniel refused his father's advice, ending up in St. Petersburg Russia, perhaps in part because his father undermined his efforts to get a position in mathematics in his native Basel, Switzerland. While in St. Petersburg, Daniel studied vibrating systems and the motion of fluids. Daniel coined the word hydrodynamics. Eventually he entered a competition for the Grand Prize of the Paris Academy (the French equivalent of the Royal Society) in 1734. Unfortunately for him, his father entered the same competition. Father Johann began what can only be described as a jealous vendetta against his own son.

After some brilliant work with the help of Leonard Euler, Daniel attempted to publish his discoveries, only to find that his publication was blocked and his ideas contained in publications by his own father. Ultimately, Daniel's work was published, but only after his father published work plagiarized from his own son, going so far as to predate the publication so as to claim priority in the discoveries. History has sided with Daniel in his long-standing feud with his father. Daniel's work was plagiarized.

The second case of blatant plagiarism is that of Elias Alsabti. Alsabti worked in medical research in the late 1970s and early 1980s, doing work at Temple University, Jefferson Medical College (Maryland), Anderson Hospital in Houston, the University of Maryland, and Boston University. Alsabti published an impressive number of research papers, but once his *modus operandi* was revealed, it became clear that none of the papers represented his own work. His scheme was to publish plagiarized work in relatively obscure medical journals, apparently in the hope that he would not get caught. While working at Anderson Hospital, Alsabti was confronted by his research adviser, Frederick Wheelock, for faking his data. Wheelock dismissed Alsabti. Apparently before leaving, however, Alsabti lifted a copy of a grant application and the drafts of several manuscripts, because in the succeeding months, work by Wheelock appeared in a number of obscure non–peer-reviewed journals across the globe. Alsabti was finally outed when Wheelock came across a paper under Alsabti's name that was virtually word for word what Wheelock had written. In the end, Alsabti published more than 70 research articles. He even lifted articles straight out of one journal and submitted them to another. When Wheelock tried to get important science journals such as *Nature*, *Science*, and the *Journal of the American Medical Association* to publish warnings against the plagiarism of Alsabti, he encountered very strong resistance. Apparently, the check and balance system was slow to respond to this embarrassing ethical lapse. Finally, the journal *Lancet* published an exposé, and Alsabti fled the country.

These two case studies show how blatantly plagiarism can occur in science, but their bizarre details illustrate why this kind of unethical behavior remains uncommon. It is the more subtle use of intellectual property that is more problematic in science. Unlike faked data and biased interpretation, there is no obvious mechanism in science to prevent plagiarism. One reason is that if good science is plagiarized, it is still good science, so it will not have the obvious flaws of other unethical behavior. One can argue that plagiarism is not harmful to human societies or even to the progress of science, because it does not release bad science to the community. Nevertheless, many individual careers have been hindered by this behavior.

Faked Data

Unlike plagiarism, whose principal impact is on individuals whose work is stolen, the faking of data can have a negative impact on the progress of science as a whole. In most cases in which scientists have faked their data (including Alsabti, by the way), the one faking the data creates bogus observations that agree with a hypothesis, which the scientist believes is, in fact, verifiable. Generally, the manipulated or created data matches this theory. Often such behavior comes about due to pressure to "publish or perish." The scientist runs out of time, money, or ability, and decides instead to cheat. This behavior involves blatant dishonesty.

It is impossible to estimate the number of cases in which data fakers have gotten away with their misdeeds. From the cases that have come to light, however, a pattern emerges. Generally, the dishonest scientist works alone. As already stated, this is a red flag for unethical science. As a rule (with notable exceptions), the faking of data is discovered when the work cannot be reproduced by other scientists.

Whereas blatant plagiarism is rare, examples of publicly revealed faking of data is surprisingly common, reaching to the highest levels of research. A few examples of faked data will illustrate this problem.

Probably the most famous example of important science including faked data comes from the great Gregor Mendel, the inventor of modern genetics. While his theories survive with great success to this day, and despite what was clearly a lot of hard work and good science on the part of Mendel, the verdict of the scientific community is unanimous. Mendel "improved" his data. According to Mendel's (correct) theory, when homozygous dominant and homozygous recessive traits are crossed, the second generation should have a 3:1 dominant to recessive phenotype ratio. The problem with Mendel's published work is that his data was perfect. The results agreed with theory to an extremely improbable degree. No biologist to this day can generate such perfect data, unless they do dozens of experiments and simply ignore the ones that are not "perfect." Mendel faked his data. The fact that he was not a well-trained scientist and that he worked in isolation were contributing factors to his bad behavior.

One hopes that the scientific community has progressed to the point that faking data would be less common. One might as well hope that human nature will change—a slim hope. A number of important examples of faked data have emerged in recent decades. First, there is the case of Mark Spector. A brilliant graduate student working under his research adviser, Efraim Racker, at Cornell University, Spector produced data implying he had discovered a protein responsible for turning on cancer cells. These events occurred in 1980–1981. Spector was almost immediately hailed as a biochemical superstar, with his adviser Racker basking in the glory. Spector showed that a particular enzyme undergoes phosphorylation by another enzyme, but that this action only happens in cancer cells. With hindsight, we now know that Spector was faking his data—applying a spike of radioactive iodine to his autoradiograms. Racker gave a lecture on the spectacular discovery to an audience of 2000 at the National Institutes of Health. Within weeks, research groups across the globe raced to reproduce Spector's results. The results were ambiguous at best. When researchers sent their samples to be analyzed by Spector, the results were positive. Finally, his fellow students and Racker confronted Spector, demanding he reproduce his results. He was not able to do so. Spector's career as a scientist came to an ignominious end.

But that is not all. There is the case of Victor Ninov, working at Lawrence Berkeley National Laboratory. In 1999 he published results, claiming to have discovered elements 116 and 118. The data he generated matched theoretical calculations for these elements. The problem came to light when other researchers were unable to reproduce the results of Ninov. The original paper included many names, but only one scientist, Ninov, had collected the data. Again, we see the problem of working in isolation. The research

group at Lawrence Laboratory published a retraction and Ninov was fired.

And there is the case of J. Hendrik Schoen. Working at Bell Labs, he claimed to have discovered a molecule that could be used for "quantum computing." If true, this would represent a very important breakthrough for computer scientists in their ability to store information in ever-smaller areas. Many considered Schoen on the fast track for the Nobel Prize. Eventually, it was discovered that Schoen took data from an earlier paper and massaged it for another paper. Some of his data was manipulated; other data was outright fabricated. The scandal was revealed in 2002. The similarity to the Ninov case is striking. There were 20 scientists on the team, getting at least some credit for the discovery, but it was revealed that the actual data was all generated by Schoen, working in isolation.

Then there is Dr. Hwang Woo-suk. His lab in Seoul, South Korea, had already claimed to be the first to clone a dog. Then they came out with the claim to have been the first to clone human stem cells. They reported cloning 11 different stem cell lines. Eventually, it was apparent that only two stem cell lines were cloned, and that the others were "doctored." In the end, it is doubtful that the research group cloned any embryos at all. Hwang was about to be appointed the head of a worldwide network of stem cell researchers. Instead, he resigned in disgrace.

Biased Interpretation of Data

It has already been claimed that blatant plagiarism is rare and that faking of data not common, but it is also not rare. These activities involve willful deceit. The problem of biased interpretation is much broader than plagiarism or faked data, for the simple reason that it does not require a blatantly dishonest unethical behavior. Liars fake their data, but honest people will publish biased research results. Bias is a fact of life for human behavior. We all tend to see what we want to see. The problem is that this perhaps relatively benign fact of human nature is completely unacceptable in the pursuit of science. One can make a strong argument that biased interpretation is the most dangerous of the ethical issues in science. Plagiarism of good science does not infect science with misleading results. Faked data is more problematic, but even in such cases, the results usually involve helping the data go in a direction it will eventually go of its own accord when other scientists do not take such unethical shortcuts. Biased interpretation is more subtle, but its effect can be more insidious for this very reason. Consider the case of medical research into new treatments. Biased interpretation, either of the efficacy of such treatments or of the dangers of side effects, can result in the loss of thousands of lives and the waste of billions of dollars. Indeed, we have documented cases of exactly this happening.

As already pointed out, nothing is more common in human behavior than finding a result we hope to find. If we are convinced that another loves us, we will irrationally fail to notice blatant evidence to the contrary. Aware of this insidious effect, the scientific community has established several norms that can prevent biased interpretation moving scientific conclusions in the wrong direction. These checks and balances include peer review, of course, but in addition, let us mention other means that are part of standard scientific operating procedure to prevent biased interpretation:

1. Using instruments rather than human beings to generate data whenever possible.
2. Proper and careful application of statistical measurements to establish valid cause-and-effect relationships.
3. Avoidance of obvious conflicts of interest.
4. Use of blind and double-blind (where appropriate) studies.

Any time a scientist records observations, there is a significant chance he or she will not produce objective data. If we expect the light to be brighter or if we anticipate a particular animal behavior, then that is what we will observe. Such observer bias has been repeatedly demonstrated by researchers under controlled circumstances. Scientists are often shocked at their own biased observations. "Surely, not I," the scientist will say. One way to avoid this is to have an instrument measure the brightness or to observe and count the animal behaviors. This approach alone does not prevent biased results, but it is a major tool. When in doubt, use instruments to collect data.

Whether or not a change in light brightness or in the number of animal behaviors is significant can perhaps be subjective. Is a 5% increase in brightness significant? The answer is that, in fact, the decision of validity is not subjective, at least in principle. As a matter of common practice, scientists always subject their data to certain statistical measurements. Lack of good statistical evaluation can be caught at the peer review level. Such standard statistical measures can establish with great certainty whether there is a valid effect on a dependent variable upon changing an independent variable. This is all well and good, except that statistics alone cannot settle the question of validity. This is true because of the possibility of hidden or ignored variables which need to be controlled. There is always the possibility that the statistical/mathematical connection between variables is correlation rather than cause and effect. For example, perhaps the residents in a particular community who are exposed to a particular toxin and who have a higher rate of leukemia may also be exposed to another substance that is the actual cause of the observed effect. Statistics must be used, they may reduce or even eliminate bias in certain cases, but they are not a silver bullet to eliminate all experimental bias.

If we ask the fox to guard the henhouse, we should not be surprised if a few hens disappear—and if they do, we need not look for an external cause of such disappearances. If we ask the creators of genetically modified crops to test whether they are safe, we should not be surprised if their result is biased toward a positive conclusion. The National Institutes of Health, the Food and Drug Administration, and other research and regulatory agencies require researchers to apply independent analysis of results to avoid biased interpretation. Unfortunately, these efforts to exclude conflict of interest are not always effective. In our first case study, we will see that those who evaluated the safety of Vioxx had a very strong conflict of interest. Good science always avoids even a hint of financial or other conflict of interest. Can this be entirely avoided? The answer is no, because in almost all science, the researcher has an expected conclusion, and we can argue that the conflict of interest is to see the expected result. Nevertheless good science avoids obvious avoidable conflicts of interest.

Blind and Double-Blind Experiments

Of all the means of avoiding biased interpretation, probably the most important is the use of blind and double-blind experimental techniques. There are actually two kinds of blind experiments. In one type, the experiment is designed so that there is a test and a "control" group. One group has the independent variable changed when the dependent variable is observed, while the other does not. The point is that the one collecting the data is not aware of which is the test group and which is the control group. Of course, if an instrument is being used to collect the data, this kind of blind study is not required. As an example of this kind of blind study, if an experiment is designed to test the effect of iron concentration in the soil on the number of tomatoes produced, the scientist would prepare two different plots, one with increased iron and the other with none added. The one collecting the tomatoes would not know which plot was which. This is a fairly crude example, but it illustrates the

technique. In this case, if the researcher expects more tomatoes with added iron, he or she will not know which field to expect the increased yield from. Whether or not this kind of blind study is needed will depend on the nature of the experiment.

A second kind of blind study is relevant only to human studies. The subjects of the experiment are not aware which group they are in. If this is a medical experiment, one group gets an actual treatment, while the other gets a fake "placebo" treatment. If we are studying the production of tomato plants, we need not be concerned whether the plants are aware of whether they are in the test or control group. The **placebo effect** does not apply to tomato plants. However, countless studies have shown the power of the placebo effect in human studies. If we take a medicine which is supposed to reduce the intensity of a headache, then our headache symptoms will decrease, even if we take a "sugar pill" (a placebo). Blind studies do not eliminate the placebo effect. Even when test patients are aware that there is a 50% chance they are receiving a fake treatment, they will experience a psychosomatic change in the expected symptom. However, when blinded, we can assure that this effect is equally applied to both test and control groups.

For both medical and psychological research, the gold standard is the double-blind study. A double-blind study is "double" blinded, in that neither the human subjects of the experiment, nor the people who are making the observations or collecting the data, are aware of which people are in the test group or the control group. Double-blind experiments are very expensive. These experiments require three categories of people. One group administers the medicine, applies the therapy, or subjects the person in the group to a particular psychological situation. The administrator *does* know who is in the test or the placebo-receiving group. A second category is the people employed to interview the patients or to monitor them for particular symptoms. Of course, the third group is the patients. All need to be paid, knowing up front, that half of the data is not even going to be used, as the treatment was a fake one. It can be difficult to enroll people in such studies. Imagine that you have a potentially fatal kind of cancer and are asked to take medicine, knowing that there is a 50% chance that it is a fake medicine, with no chance of providing a cure!

Case Studies of Biased Interpretation of Data

Let us consider some examples of well-known, documented cases of biased interpretation of data. Our first is the case of the anti-arthritis and pain medicine known as Vioxx. This drug is the best known of a class of NSAID (nonsteroidal anti-inflammatory drugs) compounds. It was put on the market by Merck in 1999 and touted as a miracle drug, especially to treat chronic arthritis. By 2003 it became one of the leading all-time sellers, reaching $2.6 billion in sales worldwide. When it came to its promises of helping chronic arthritis sufferers, the drug fulfilled expectations.

However, trouble followed this miracle drug. When Merck published data, and when they informed the FDC about such data, they failed to include a group of studies that showed evidence of a significant increase in strokes and heart disease. After the fact, Merck scientists defended themselves by noting that they did not suppress such data. In fact, this is true. They published all their studies, but chose not to emphasize certain studies in their final reports on the safety of the drug. It just so happens that the studies they did *not* emphasize included those showing a significant incidence of fatal heart disease. Merck further defended itself by noting that the negative results were from smaller studies or from comparison studies to other NSAIDs, rather than studies versus a placebo. Such arguments will probably not convince many outsiders. This is a rather obvious case of scientists choosing to ignore data that they did not like. In the end, many thousands of patients died of heart disease. Under pressure, Merck pulled the drug

from the market in September 2004. Merck will end up spending many billions of dollars settling lawsuits over this ethical disaster.

What would have prevented this disastrous result? All studies were double blind. The answer in this particular case is that the chief means of preventing biased interpretation—which was not applied—was avoidance of conflict of interest. The FDC has changed its policies to reflect this problem.

A second case of biased interpretation of data is the cold-fusion debacle. Two scientists from the University of Utah, Stanley Pons and Martin Fleischmann, made the stunning announcement of the discovery of cold fusion at a press conference on March 23, 1989. They reported what would be a truly world-changing discovery if true: that fusion of deuterium atoms to produce helium can occur using a palladium catalyst at room temperature, potentially producing a vast amount of energy. Within a few weeks, Pons and Fleischman were on the cover of *Newsweek*, holding their fusion devices in their hands. Such a discovery would seem to violate some common-sense notions of the difficulty of joining positively charged nuclei. Theories predict (and prior experience confirms) that successful fusion collisions of atoms require millions of atmospheres of pressure and a temperature of millions of degrees. Such conditions are obtained at the center of stars, but are fantastically difficult to produce in a laboratory. The Utah chemists reported the production of energy from their electrochemical cells far in excess of what could be produced by a normal chemical reaction.

Within weeks, hundreds of scientists were attempting to reproduce the results of Pons and Fleischmann. At first, some positive results were achieved. Some got excess heat, others saw the required neutrons or the gamma radiation reported by Pons and Fleischmann. However, none of the attempts to reproduce the work at the University of Utah showed all three pieces of data. Within just a few months, questions and doubts mounted, with the result that by November 1989 the U.S. Department of Energy produced a report concluding there was no convincing evidence for cold fusion from any of the studies done up to that time. Pons and Fleischmann have still not conceded defeat on the matter, but we can assume that this is a case of bad science and biased interpretation.

What happened? Pons and Fleischmann were working in relative isolation. Add to this, they were making discoveries of neutrons and gamma particles when they had no expertise in such products of high-energy nuclear reactions. Their announcement of their results at a news conference raises red flags as well. In the end, we can conclude that once these eminent scientists concluded that they had made a game-changing discovery, wishful thinking overcame common sense. They had discovered evidence which, with hindsight, was probably just "noise." They over-interpreted their data, and in their enthusiasm, failed to spin alternative hypotheses to explain their anomalous data. All of these are lessons for scientists in how NOT to do science.

Ethics: External Issues

Let us now consider a completely different kind of ethical issue. We will move from discussing the ethics of how science should be conducted to the ethical implications of what scientific discoveries have allowed us to do. Few will question that science has given us technologies that are wonderfully beneficial to people. Science has allowed us to communicate over great distances almost instantaneously. It has given us the ability to explore the universe and has led to finding cures for many deadly diseases. No doubt some of these technologies have ethical pros far outweighing the cons. With this as a background, we should acknowledge that scientific technologies have given us the ability to do things that most or all people would find to be very troubling from an ethical cost/benefit analysis. We now have the ability,

using scientific discoveries, to destroy all life on the planet. Our job here is not to discuss the ethical pros and cons of stem cell research, genetically modified crops, nanotechnology, human cloning, and eugenics. Rather, it is to provide the basis and ask the questions we can use to frame the debate.

Is science ethical? This really is a nonsense question. Science is a method and a way of thinking. This method and way of thinking is neither ethical nor unethical. Knowledge gained by scientists is neither good nor evil. Knowledge of the workings of the physical world is marvelous. It is beautiful, but it is not good or bad. It is when human beings apply their knowledge of how the physical world works to create technologies that ethical issues are raised. Visionary philosopher Francis Bacon said, "Nature, to be commanded, must be obeyed." In other words, if we want to make nature do things she normally would not do, we must follow her rules. Once we know how genetic information is reproduced and expressed, we can use that knowledge to manipulate genetic information to create "unnatural" plants and animals. This, of course, raises ethical issues.

In order to begin discussing ethical implications raised by scientific discovery, it will be useful to distinguish two broad categories of scientific research: pure and applied. "Pure" research or pure science is the study of the laws of nature. It is sometimes called basic science or basic research. It is the search for the underlying cause-and-effect relationships that determine what happens in the universe. The distinction is not completely clear cut, but in pure scientific research there is no "purpose," other than to plumb the workings of nature. This research generally is done at research universities and at government research laboratories. Scientists in such institutions often must justify their research grant proposals based on possible applications, but if one talks to these researchers, one will discover that for them it is not really about the applications but about the search for knowledge about how nature works.

The other kind of scientific effort is applied science. Those doing applied research are working to produce new technologies for practical applications. Their methodology is somewhat different than that of pure science. Often it involves trial and error and engineering. Such research is done mainly by corporations, but also by government laboratories. There is money to be made from applied research, so we should not be surprised that more scientists are involved in applied rather than pure science.

The distinction just made is useful for us, because generally pure research brings up few—if any—ethical implications. It is in applied research—when scientists try to use their knowledge of how nature works to make nature do things she normally would not do—that ethical issues come up. When Einstein discovered that $E = mc^2$, we do not find ourselves asking whether it was unethical to discover that matter can be turned into energy. On the other hand, when Einstein, along with several fellow physicists, signed off on a letter suggesting to President Franklin D. Roosevelt that the United States build a nuclear bomb, surely there were major ethical ramifications to this applied research.

Are there types of research that should not be done, and categories of technologies that should not be created at all? Should legal sanctions be set up, prohibiting certain research programs? What about technologies that should be pursued, but should be studied extremely carefully and subjected to rigorous analysis before such technologies begin to be used or sold? It is not our job here to inform students which scientific research programs are ethical and which are not. Our goal is to provide the tools and to ask the questions that can be used by any human being to make ethical decisions about which kinds of technologies ought to be pursued.

A Few Things to Think About

Consider a few points that offer useful perspective in our attempts to ask ethical questions about scientifically produced technologies:

1. There are scientific technologies that you believe are a great idea ethically today, but 30 years from now, you will think they were a large ethical error to pursue.
2. There are scientific technologies that trouble you ethically today—that you think should not be pursued, but which you will think are just fine 30 years from now.
3. The pace of new scientific discoveries and the speed with which new technologies are coming online has become so rapid, that our ability to grasp the ethical ramifications cannot keep up with the research.
4. Nonscientists should not completely trust decisions about what research programs to pursue to scientists because they are biased. Their attitude is "let's do the research" and worry about the ethical implications later.

Presumably, morality does not change, assuming that there is such a thing as moral absolutes. However, our perception of the ethical pros and cons of scientific technologies will change with time. In many cases, we only see the possible benefits and do not anticipate the possible negative consequences of a new technology. What seems patently obvious today may, upon experience, look dubious. The history of scientific discoveries is replete with examples of what is known as the law of unintended consequences. We take Vioxx to reduce arthritic pain, but end up dying of a heart attack. We introduce the cane toad in Australia to control the greyback cane beetle. The result literally came back to bite us, overturning entire ecosystems and driving several species of snakes to extinction.

The story of the pesticide DDT provides a good example of a technology that was universally accepted as an ethical good at first, but eventually turned out to be an unmitigated disaster. In 1939 the pesticide properties of this compound were discovered. By the end of World War II, DDT was used to prevent mosquito-borne diseases, especially malaria. This chlorinated pesticide was thought to be the ideal disease-preventing compound because of its persistence (resistance to breakdown in the environment). Literally millions of lives were saved by the widespread spraying of this miracle pesticide. However, it turned out that one of the perceived advantages of DDT—its persistence—was the reason it turned out to be an environmental disaster. DDT is an insidious toxin whose concentration is amplified as it is passed up the food chain. In 1962 Rachel Carson published the book *Silent Spring*, which became the opening salvo for the environmental movement. Carson noticed that in areas sprayed with DDT for mosquito control, birds had virtually disappeared—thus the name *Silent Spring*. DDT caused the shells of eggs to become thin, dropping birth rates of birds to near zero, especially for birds of prey. The bald eagle was pushed to the edge of extinction. By the late 1960s, DDT was phased out, and by 1975 it was banned for use in the United States. An ethical miracle had become an ethical disaster.

Are genetically modified crops the next miracle technology, or will they turn out to be an unmitigated environmental and human disaster? What about food irradiation or nanotechnology? Only time will tell.

An example of the second kind of change over time in our ethical perspective is provided by *in vitro* fertilization. When IVF (children conceived this way are sometimes known as test-tube babies) was introduced, it was declared unethical or immoral by most religious groups. The great majority of people rejected this technology as ethically unacceptable, despite the fact that it was a means for otherwise infertile couples to have children. There was talk of banning the procedure. The world's oldest test-tube baby, Louise Brown, is now 32 years old. Today, if infertile couples want to seek funding for expensive IVF treatments, they can often get financial help from the same churches that once declared the procedure unethical. What changed? People got used to the idea and, over time, they began to notice the ethical good from IVF. This "unnatural" technology gave loving couples access to the greatest of human blessings.

Today, human cloning is almost universally rejected as "unnatural" and unethical. It has been banned in many countries, including the United States. Might human cloning be the IVF of the future, once we get used to the idea? Again, time will tell.

Fifty years ago, the pace of scientific discovery was much slower than it is today. Generally, new technologies came online 10 years, 20, 30 or more years after they were conceived of in theory. By then, nonscientists who cared to consider the issues had sufficient time to reflect on the ethical pros and cons. This did not eliminate ethical blunders, but it helped. All this has changed. Advances in science are produced at a furious rate. Many radically new technologies are produced by applied science and are brought to the market before the scientifically uninitiated are even aware of the technology, never mind the ethical implications. Genetically modified crops (GMCs), sometimes known as genetically modified foods, are a good example of this ethical quandary. Crops with new traits are created by using a viral vector to introduce genes from bacteria, animals, or other plants into the seeds of food crops such as corn, soybeans, and tomatoes to produce plants with desirable new traits such as pest or herbicide resistance, greater nutrition, and drought or salt resistance. The potential ethical good is obvious. The problem with GMCs is that the potential cons are just that—potential rather than actual, at least for now. Problems such as cross-pollination to produce super-weeds have not yet been observed. The "cons" of GMCs are principally what might happen, not what has already happened. It has been estimated that 75% of all foods eaten in the United States contain genetically modified foods, yet the average American is not even aware of the possible dangers. We would do well to remember the law of unintended consequences. GMCs are banned in the European Union. This is a clear example of the rate of scientific discovery outpacing our understanding of the possible ethical issues.

"Nonscientists should not completely trust decisions about what research programs to pursue because they are biased. Their attitude is 'let's do the research' and worry about the ethical implications later." Some might consider this a radical statement. Consider the fact that this is coming from a scientist. It is not that scientists are not ethical people or that they do not care about the ethical implications of the research they conduct. We already know that some scientists have behaved unethically, but my experience is that, on average, scientists are a cut above the norm when it comes to concern about ethics. The problem is not that scientists do not care, but that they have a conflict of interest. Their "interest" is to do what scientists love to do—to investigate nature. Most scientists want to pursue the knowledge, wherever it leads. They care about ethics, but they may care about their research even more.

Here is why this should be a concern. The attitude of scientists tends to be "we can deal with the ethical ramifications after the fact." " If it is revealed that there are serious ethical problems induced by the new technology, we can stop using the technology or ask scientists to solve the issue. The problem with this is that once we let the genie out of the bottle, it is difficult to put it back inside. Once a genetically modified crop is put out into the market and corporations are earning billions of dollars from such crops, ethical concerns may no longer win the day. Even if they do, once a dangerous gene is accidentally leaked out into the environment, it might be literally impossible to take it back. To use a different example, once the nanoparticles escape into the environment, it may be far more difficult to undo the damage than it would have been to prevent the damage in the first place, and

we may find profit-motivated corporations resisting such efforts.

If this is true, then it seems a wise policy to not leave ethically related questions about what kinds of research ought to be pursued in the hands of scientists alone. Well-informed nonscientists should be looking over the shoulders of scientists. It is not exactly a case of the fox guarding the henhouse: it is more like the fox guarding the fox house. We might want some hens watching the fox house. As a scientist, please permit me to make an appeal to nonscientists. Be involved, ask questions, and do not be afraid to speak your mind, and perhaps even pursue civil action if you cannot be heard.

The Precautionary Principle

Let us assume for a moment that we have wise mechanisms in place, which allow both scientists and nonscientists to have input into what technologies will be pursued, and which ones will actually enter the market. This, of course, is a big assumption. How will we then come to a consensus on which ideas to move forward? What will be the principles upon which we make such decisions? Again, it is not my intention to provide the answers here. However, one potential standard for such decisions has been proposed. It is known as the **precautionary principle**. This principle states that, *unless and until it has been demonstrated beyond a reasonable doubt that a particular technology is not harmful to humans or to the environment, such technologies will not be permitted to come into general use*. In other words, under this principle, the burden of proof is not with those trying to caution against new technologies, but with those trying to introduce them. It is sometimes called the "better safe than sorry" principle. Be aware that this is an extremely high ethical standard, which, if applied vigorously, would have prevented the introduction of many technologies that we all love to use. On the other hand, its supporters will note that, given the great number of well-documented examples of the law of unintended consequences, it is the only reasonable standard for making ethical decisions. It would be fair to say that those pushing this standard are more than a little skeptical of the ethical intentions of multinational corporations.

The possible pros of using this principle are great. Had it been applied, then the tragic environmental disaster caused by the use of DDT would have been prevented. It also would have caused us to avoid the tens of thousands of deaths from the use of Vioxx.

Further having said this, the possible negative consequences of a vigorous application of the pre-cautionary principle bear mentioning. If we require proof beyond a doubt of no harmful effects, then the rate of introduction of new medical treatments would presumably drop precipitously. The next H1N1-like epidemic might be worse than the last, but with the added problem that no vaccine would be available. How can we prove that the vaccine will do no harm?

It is nearly impossible to prove that a new technology will do no harm. Pro-business groups will resist this principle very strongly. They will argue that economic growth would be stifled and millions would lose their jobs. The proponents of the PP will have to respond to this legitimate criticism.

It is worth noting that in the European Union, the precautionary principle is found in many signed agreements. In the United States, those who put genetically modified foods onto the market are merely required to show that no known harm has yet been identified. This is a rather low standard of evidence. In Europe, the precautionary principle is applied, with the result that GMCs are completely banned, at least for now.

Rational people will conclude that the application of the precautionary principle should be pursued with some sort of balance, weighing the possible benefits, and not just the possible harm of new technologies. Where should this balance lie? To quote a famous author, *that* is the question.

How We Know

The Experimenter and the Experiment

By Martin Goldstein and Inge F. Goldstein

The Uncertainty Principle

There is a principle in physics discovered in this century called the uncertainty principle. It states a limitation on our ability to measure anything we want to, with any accuracy we choose.[1]

Suppose we have an object, a bullet, for example, known to be moving in a certain direction with a constant speed, and we want to find out two things about it at a certain time: (1) where it is at that moment and (2) how fast it is moving at that moment. The experimental procedure for doing this is simple to understand—we might imagine using high-speed photography, photographing the bullet with two flashes of light a known short time interval apart. The first photograph is sufficient to answer the first question; the distance traveled by the bullet between the two exposures answers the second.

Electrons can also travel through empty space at a constant speed, as the tube of a television set demonstrates, and we might imagine asking the same two questions about the moving electron that we did about the moving bullet. Indeed, the experiment can be tried in much the same way, using photographic film to detect the result of two flashes of light impinging on the moving electron.

However, we find ourselves in difficulties. We find that because of the small size of the electron the accuracy with which we can locate its position from the first light flash depends on the energy of the light used: the higher the energy of the light, the better the accuracy. But the light, in the process of being reflected from the electron, gives some of its energy to the electron, and the higher the energy of the light used, the more the electron is affected. The result is that while the first flash of light can locate the position of the electron to within certain limits set by the energy of the light, the electron's speed has been so perturbed by the first flash that the position of the electron on the second flash no longer reflects its original speed. The more accurately we try to determine the position, the higher energy of light we must use, and in turn the less we know about what the electron's speed was before the experiment. Light has a large effect on the electron but not on the bullet because the electron is very small and doesn't weigh much, while the bullet is heavy. We find we cannot know the *speed* of the electron accurately once we have measured its *position* accurately.

Thus the result of our experiment is accurate knowledge of where the electron is, combined with poor knowledge of how fast it is moving. Now there are other, different experiments we could perform to gain accurate knowledge of the electron's speed. The details of how we might do this do not matter: what matters is that the very measurement that tells us the *speed* accurately deprives us of knowledge of the *position* of the electron. The more accurately we learn how fast it is moving, the less accurately we know where it is.

So we are forced to choose; we cannot know both position and speed. The two kinds of knowledge are mutually exclusive; the better we know one, the worse we know the other. And modern physics tells us there is no way out of this dilemma. It is a fundamental law of nature. *The very act of measuring changes the system in unpredictable ways.*

A Useful Metaphor

The uncertainty principle is a useful metaphor for a problem that affects all of science to a greater or lesser degree. When we do an experiment, we tend to experiment is isolated from any influences not under our control, that the instruments used measure and the experimenter making the observations are somehow "outside" the thing being studied, and not affecting it. In fact, such an idealization is often wrong. The person doing an experiment is part of the experiment. He may be affecting the outcome by his presence, by his preconceptions, by the technique of the experiment itself.

Sometimes the difficulty lies not in the way the experimenter influences the system being studied but rather within his own mind. What he sees is influenced by his expectations, and he reads into the situation features that are not really there. Sometimes it is a simple matter of tending to make those errors in recording data that support his preconceived idea of what the results should be.

Whatever form the problem takes, it is widespread in science. It crops up in unexpected ways, and we must be alert to it.

The Smart Mice

An assistant of the Russian psychologist Pavlov once published some remarkable results showing the inheritance of acquired characteristics. He had found that when mice were trained to solve mazes, their children could solve them faster, and their grandchildren faster still. The effect continued to the fifth generation. The first generation required approximately 300 trials to solve the maze; successive generations required approximately 100, 30, 10, and 5 trials.

Pavlov later repudiated these results; apparently the assistant, over time, had unconsciously learned to train mice better how to solve mazes. The fifth generation of mice weren't really any smarter than the first. It was the experimenter who had changed[2,3]

Placebo Pills In Drug Trials

In several of our case histories, we have discussed the concept of a controlled experiment: the idea of taking two groups of subjects or cases identical in all relevant features except one, the one whose effect on the experimental group we wish to study.

For example, we wish to test a certain drug for its effectiveness in relieving cold symptoms. We take a large number of people with colds, divide them at random into two groups, and give one group the drug. It would seem reasonable that if the group given the drug suffers less and recovers faster we have proved the value of the drug. However, this isn't so. Trials of drugs are not usually made this way anymore, because the people given a drug will usually feel better, or *think* they feel better. They have been given a medicine and they will expect it to work; this expectation is enough to make them think they have been helped.

The two groups in the experiment were supposed to differ in only one way: one group has received the drug and one group has not. But in fact they differ in two factors, not one, a situation we may represent in Table V.

The difference in results of drug tests may often be due to factor (2) rather than factor (1). So where possible in drug trials the control group is given some harmless and inactive pills—sugar pills, for example—so that the two groups will not differ by factor (2).

For many years, doctors have dealt with patients with whom nothing discernible is wrong, yet who complain of various illnesses. They often find honesty with the patient does not pay—telling a patient there is nothing wrong with him sometimes leaves him dissatisfied and mistrustful. It has been found easier if not more ethical to prescribe some harmless medication. Often the patient feels better—he might have anyway—and at least he feels that the doctor has taken him seriously. A "medicine" of this type is called "placebo" from the Latin word *placare,* "to please" (the word *placate* comes from this source also). The curative effect of the inactive pill in the "controlled" experiment, when it has no real curative power, is called the "placebo effect."

Obviously, in an experiment on mice, which make no distinction between medicine and food, it is not necessary to use a placebo to avoid this particular source of error.

Blind and Double Blind

There are other sources of error that might arise in such an experiment even if a placebo is given to the control group. In most tests of drugs it is not sufficient to rely on the patient's subjective feelings as to whether he is better or not. More often a physician

Table V Group Differences

Test group	control group
1. Has received a drug	1. Has not received a drug
2. Knows it has received a drug	2. Knows it has not received a drug

Table VI Group Differences

Test group	Control group
1. Has received a drug	1. Has not received a drug
2. Symptoms evaluated by a physician who knows the patients have received a new drug	2. Symptoms evaluated by a physician who knows the patients have not received a new drug

is needed to evaluate by professional criteria whether improvement has occurred in one group and not the other. When the evaluation requires some simple objective test, such as measuring the amount of sugar in the urine, there is not much of a problem. But often the evaluation of improvement in a patient's disorder requires a subjective judgment on the part of a physician—for instance, have the schizophrenic patients who received drug A improved more than the control group who received a placebo?

Such an evaluation requires paying attention to many details of a patient's behavior and balancing seemingly contradictory information to obtain a single judgment: yes, there has been improvement; no, there has not. In such a delicate procedure there is a risk that the physician's bias may cause him unconsciously to judge patients whom he knows have received a new experimental drug differently from those whom he knows have received a placebo. If he is personally biased in favor of the drug, he may lean one way; if he is biased against it—prefers some tried and tested drug he has had success with to this new and dubious one—he may lean the other.

Again, the two groups differ in more than one factor (Table VI). Again, a difference between the results of the tests on the two groups may be due to factor (2). Thus it is preferable to have not only the patients but also the evaluating physicians be ignorant of who has received the drug and who has not. Such an experiment is called a "double-blind" experiment.

Unfortunately, it is not always possible to keep patients and evaluating physicians ignorant of the treatment used. For example, if we wish to compare the relative effectiveness of surgery and drug therapy for some form of cancer, both patient and doctor will know. However, when the double-blind design can be used, it is a better procedure. It was used in the experiments to evaluate possible genetic factors in schizophrenia where the psychiatrists who had the job of diagnosing whether the children or parents of schizophrenics had a greater-than-normal chance of being themselves schizophrenic were kept in ignorance of whether the individuals they examined were or were not blood relatives of schizophrenics.

The Lively Flatworms

The way in which the expectations or biases of an experimenter can influence the results he thinks he observes has been studied extensively by the psychologist Robert Rosenthal[4] He describes one experiment, for example, where undergraduate biology students were asked to observe how many times members of a species of flatworms turned their heads or contracted their bodies. The students were led to believe that one group of worms was expected to move a lot, while a second group was expected to move only a little. In fact, the two groups of worms were identical. "[The] observers reported twice as many head turns and three times as many body contractions when their expectation was for high rates of response as when their expectation was for low rates of response."

Mental Telepathy

The results of the above experiment depended on the fact that the experimenter had to make a subjective evaluation of whether a worm did or did not, at a *certain time, move its head*. But the problem is present even for experimental operations that do not seem to depend on subjective judgments; for example: recording the results of experimental observations. Clerical errors will always occur when results have to be written down. But, as is well known to anyone who is in the habit of checking the addition of salespeople in stores, the errors do not occur at random. In experiments on mental telepathy, where one person, the subject, had to guess which of several symbols a second person was concentrating on and attempting to transmit telepathically, a third person recording the guesses of the subject *and knowing which symbol the second subject was "transmitting"* tended to make errors in recording the result that depended on whether he himself believed in mental telepathy or not. If he was a believer, the errors in recording tended to increase the score of the guesser; an unbeliever tended to make the opposite kind of error.

The Clever Horse

Another example given by Rosenthal is the following[4]:

Probably the best-known and most instructive case of experimenter expectancy effects is that of Clever Hans [studied by the German psychologist Pfungst]. Hans, it will be remembered, was the horse of Mr. von Osten, a German mathematics teacher. By means of tapping his foot, Hans was able to add, subtract, multiply, and divide. Hans could spell, read, and solve problems of musical harmony. To be sure, there were other clever animals at the time, and Pfungst tells about them. There was "Rosa," the mare of Berlin, who performed similar feats in vaudeville, and there was the dog of Utrecht, and the reading pig of Virginia. All these other clever animals were highly trained performers who were, of course, intentionally cued by their trainers.

Mr. von Osten, however, did not profit from his animal's talent, nor did it seem at all likely that he was attempting to perpetrate a fraud. He swore he did not cue the animal, and he permitted other people to question and test the horse even without his being present. Pfungst and his famous colleague, Stumpf, undertook a program of systematic research to discover the secret of Hans' talents. Among the first discoveries made was that if the horse could not see the questioner, Hans was not clever at all. Similarly, if the questioner did not himself know the answer to the question, Hans could not answer it either. Still, Hans was able to answer Pfungst's questions as long as the investigator was present and visible. Pfungst reasoned that the questioner might in some way be signaling to Hans when to begin and when to stop tapping his hoof. A forward inclination of the head of the questioner would start Hans tapping, Pfungst observed. He tried then to incline his head forward without asking a question and discovered that this was sufficient to start Hans' tapping. As the experimenter straightened up, Hans would stop tapping. Pfungst then tried to get Hans to stop tapping by using very slight upward motions of the head. He found that even the raising of his eyebrows was sufficient. Even the dilation of the questioner's nostrils was a cue for Hans to stop tapping.

When a questioner bent forward more, the horse would tap faster. This added to the reputation of Hans as brilliant. That is, when a large number of taps was the correct response, Hans would tap very, very rapidly until he approached the region of correctness, and then he began to slow down. It was found that questioners typically bent forward more when the answer was a long one, gradually straightening up as Hans got closer to the correct number. …

Pfungst himself then played the part of Hans, tapping out responses to questions with his hand. Of 25 questioners, 23 unwittingly cued Pfungst as to when to stop tapping in order to give a correct response. None of the questioners (males and females of all ages and occupations) knew the intent of the experiment. When errors occurred, they were usually only a single tap from being correct. The subjects of this study, including an experienced psychologist, were unable to discover that they were unintentionally emitting cues.

Section V
Pseudoscience

The Marks of Pseudoscience

By John Oakes

In the summer of 1994, Dr. John Hagelin of the Physics Department at Maharishi International University and a group of his fellow "researchers" called a press conference to announce the results of their experiments on the "Maharishi Effect." Their hypothesis: if sufficient practitioners of a type of transcendental meditation known as TM-Sidhi were to practice their technique in a geographic area, this would have a calming effect on the "collective consciousness" of the entire area, sufficient to lower the rate of violent crime in that location.

To this end, more than four thousand practitioners of TM-Sidhi descended on the Washington, D.C., area in the summer of 1993. In two-week shifts over a six-week period, the participants performed the meditation technique twice a day throughout the city. Hagelin and his associates claimed the experiment would be a "scientific demonstration that will provide proof of a united superstring field." Exactly how they proposed to connect the highly speculative superstring theory to religious pantheism and a Hindu-inspired meditational technique—and how this would be connected to the crime rate—was not exactly clear. They proposed to compare a statistical prediction of the rate of certain crimes with the actual crime experienced during the six-week period.

By now, one hopes the reader's skepticism meter is rising. Is this a publicity stunt, or a legitimate scientific experiment? Let's go back to the story. One year after the experiment, Hagelin and his colleagues invited the media to the public announcement of their findings and their analysis of the data. They reported an 18 percent reduction in the predicted rate of violent crime over the six-week period in question.

Unfortunately for those putting on the press conference, a group from the *Skeptical Inquirer* were at the meeting. These included Robert Park, future author of *Voodoo Science*. This group is dedicated to debunking pseudoscientific claims. A reporter from the *Washington Post* asked Hagelin about the 18 percent decrease in crime rates, asking, "Compared to what?" "How could you know what the rate would have been?" Obviously skeptical of the data interpretation, Park asked Hagelin about the potential bias of the "independent scientific review board" that reviewed the data. Hagelin conceded, saying, "Some of the members of the review board have had previous experience with TM." When pressed, it turned out that all were followers of the founder of TM, the guru Maharishi Mahesh Yogi.

In his treatment of the work of Hagelin *et al.* published in his book *Voodoo Science* (as well as in the *Skeptical Inquirer*), Park dismissed the study as a "clinic in data manipulation." He accused the researchers of scientific misconduct. The tone of Park's response is sarcastic and dismissive. In a published response, one of the coauthors on the original study accused Park's analysis of being "superficial, highly polemical" and "willfully misleading."

One year later, in 1995, Hagelin and his associates founded the Natural Law Party in the United States, which put Hagelin on the ballot in every state in the 1996 presidential election. The Natural Law Party promised to revitalize America based on "scientifically proven solutions," such as TM-Sidhi.

Who is right here? Is the work of Hagelin *et al.* pseudoscience, as Park accuses? Is the response of Park and friends biased and highly polemical? The answer is yes on both counts. Despite attempts by its authors to defend this study, we can be highly confident the claims being made for TM-Sidhi are a classic case of pseudoscience. How can we be sure? Because this work is studded with what we will call the "marks of pseudoscience," as we will see below. Is Park playing fair here? His conclusion in this case is correct. The work of those trying to prove the Maharishi Effect is rather obvious pseudoscience, but Park's approach is combative, and is certainly not conducive to facilitating good communication between the combatants in this battle. He calls the Maharishi Effect "palpably untrue." In other words, according to Park's worldview, it is obviously not true, regardless of any attempts to prove or disprove the effect. This is clearly not scientific behavior. What we have here is a classic case of failure to communicate.

What are the marks of pseudoscience in our case study? There are many. A list of such "marks," with definitions, will be supplied below, but for now let us note that in this case we have more than enough of these signs to make us suspicious.

1. For one, the researchers went to the media first, rather than publish their findings in respected scientific journals.
2. Then there is evidence here of what we will call the "filtering of data." We can and will question the statistical analysis of the data in question, but it is worth noting that during the six weeks in question there was a dramatic rise in the homicide rate in Washington, D.C., over that expected from similar periods. In fact, these were some of the worst six weeks for killings in the history of the city. If the decrease in aggravated assault is evidence of the Maharishi Effect, then is the increase in homicides during the same period not evidence against the effect? The researchers attempt to dismiss this fact, but are not convincing here at all.
3. In addition, there is a clear indication of what we will label "abuse of statistics." If we look closely, the violent crime rate over the six-week period was, in fact, not lower than a yearlong average. The rate of crime in the six-week study period was above average, not below average! It was lower than a calculated higher-than-normal crime rate for those six weeks, taking into account the predicted effect of the heat wave, which hit Washington, D.C., during those six weeks, and other supposed effects on crime rate, including fluctuations in the earth's magnetic field. It seems reasonable to ask whether it has been scientifically proven that the magnetic field of the earth does indeed affect the crime rate.
4. The researchers, as already noted, have a clear bias, given that all of them are employed by or strongly committed to the Maharishi religious movement.
5. Part of Hagelin's defense of his Maharishi Effect science includes a claim of being suppressed by members of the *Skeptical Inquirer*. Claims of suppression are a common trait of pseudoscience.
6. Another pattern we can detect that might raise our suspicions is that those who support the belief the Maharishi Effect is scientific are proposing a scientific revolution—a new paradigm.
7. Also, the work of these gentlemen includes what we will call argument by analogy. Whether superstring theory will prove to be legitimate science in the long run remains to be seen, but in claiming cosmic consciousness is legitimate science, the Maharishi Effect supporters also maintain this is a legitimate example of an application of stringtheory. This amounts to use of analogy rather than legitimate application of the theory.

8. Last, one can detect what we will call the "refusal to revise, even when proven wrong." It would be fair to say the Washington, D.C., study does not disprove the conclusion of the authors. One gets the very strong sense, however, that they had absolutely no intention of being affected in either direction by their "research." In other words, those doing this research have a philosophical and religious commitment to their faith, which we can predict will be absolutely impervious to the kind of evidence they are proposing to collect. Even if the experiment had produced absolutely no statistical support for their belief, the members of this research committee would not have given up their belief. This is not a scientific attitude toward evidence. When a researcher reaches a conclusion, then attempts to "prove" what he or she already believes for philosophical or other reasons, the conclusions are foreordained and not scientific. Arguably, the attitude of attempting to prove, rather to scientifically test one's hypothesis, is the chief mark of pseudoscience.

Is Science "True" and Pseudoscience "Not True"?

Having begun our study of pseudoscience with a case study, let us now consider a definition of pseudoscience.

> **Pseudoscience**—*A claim of a scientific nature (in other words, about the laws of nature) that is not supported by well-documented, reproducible scientific evidence.*

Using this definition, if a claim is not directly related to the physical laws of nature, then it cannot be pseudoscience. Most artistic, philosophical, sociological, religious, and other claims are not legitimately describable as pseudoscience.

Nevertheless, there are many claims out there we can propose labeling as pseudoscience: biorhythms, numerology, astrology, UFOs, parapsychology (clairvoyance, telepathy, etc.), out-of-body experiences, herbal medicines, pyramid power, scientology, homeopathy, chiropractic, acupuncture, alternative medicines in general, faith healing, levitation, magnet therapy, young earth creationism, ancient astronauts, dousing, etc. The list of claims that are blatant pseudoscience—or at best, borderline cases—is almost unending.

The word "pseudoscience" is, in and of itself, a strong term. The prefix *pseudo* means fake. To accuse work of being pseudoscience is to accuse the one producing it of scientific misconduct. We ought to be cautious about throwing around such potentially combative words. Another note of caution is appropriate. **Just because a claim is not scientific does not mean,** *a priori,* **that it is not true.** It is safe to say that a good majority of the areas listed as pseudoscience above are sheer nonsense. Good science, however, does not reject a proposal without study. As our friend Dr. John Hagelin would point out, much scientific reaction to things labeled as pseudoscience has itself been unscientific. Once a particular claim achieves the dubious distinction of being labeled pseudoscience by the scientific community, it becomes nearly impossible to study such questions. Scientists circle the wagons. Anyone in the inner circle of science showing even the least amount of open-mindedness toward claims labeled as pseudoscience is in danger of being publicly ostracized. Funding becomes completely unavailable. Opportunities to publish legitimate work, rare as it may be, are blocked with a kind of religious zeal.

Perhaps eventually one or two of these ideas we have labeled pseudoscience will show themselves to be scientific. The vast majority of claims for herbal medicines have never been the subject of rigorous scientific test. Therefore, by definition, they are not scientific. They only become pseudoscientific when supporters of such claims attempt to label them as scientific.

Willow bark tea was used for pain by indigenous groups for thousands of years. We now know the bark contains salicylic acid, the precursor to acetylsalicylic acid, also known as aspirin. Presumably, most herbal remedies are helping no one, except perhaps because of the placebo effect. The only ones gaining are those reaping profit from the sales of such untested treatments. But experience tells us a small fraction of these remedies will prove legitimate if tested scientifically. Fifty years ago global warming was not strictly scientific. It could have been labeled pseudoscience. Today it is definitely mainstream science. Continental drift was proposed by Alfred Wegener in 1915. At the time it had many marks of pseudoscience, yet it is now one of the two principle paradigms of geology. What changed? Continental drift changed from nonscience or pseudoscience when sufficient evidence emerged, and when a reasonable mechanism was proposed.

It seems reasonable to ask at this point why, in a modern, supposedly scientific, rational culture, there is such widespread acceptance of the kinds of claims listed above as scientific. Many of these beliefs are clearly not rational. Consider a few of the reasons for the preponderance of such unscientific beliefs:

1. Most people are not sufficiently trained in how to think skeptically and do not know how to distinguish real science from pseudoscience.
2. Human beings always need hope, especially when they face chronic or terminal diseases. When traditional, Western medicine offers little or no hope, people become susceptible to unscientific remedies.
3. The majority of people believe that science is not all there is. Belief in spiritual or metaphysical realities, whether they are real or not, contributes to acceptance of nonscientific claims.
4. Money. The greatest reason for much pseudoscience gaining such broad acceptance is because people who stand to make a lot of money have a stake in creating the false impression that such claims are scientific. By doing so they increase sales of whatever they are peddling.
5. Perhaps some of these are actually true, even if they have not yet been supported by good science. Perhaps the claim is true, but will they not ever be supported by scientific evidence? We will end this chapter with a discussion of such potential "borderline cases."

It has already been stated that just because something is not scientific does not necessarily mean it is not true. Many of those who accept these unscientific claims are quite sincere in their beliefs. If that is the case, then why do scientists react so strongly to unscientific claims masquerading as science? Why should the nonscientist be concerned with hunting down and identifying claims as pseudoscience? A good answer is that attaching the word "scientific" to a claim in our culture has a very powerful effect. Even if we allow for the possibility a small minority of these unscientific claims might one day prove to be scientific, this does not change the fact that many in our society have been deceived into thinking they are scientific when that is not the case. Millions have rejected scientifically supported medical treatment to pursue unscientific treatment. Perhaps some have been helped by doing this, but thousands have died as a result. If nothing else, we can say without a doubt that tens of billions of dollars have been wasted chasing pipe dreams.

Marks of Pseudoscience

The goal of this section is to greatly increase your ability in the art of detecting pseudoscience. There are a number of characteristics common to pseudoscientific claims, which we will call "marks of pseudoscience." Sometimes it seems as though pseudoscientists have received their training at conferences designed for the purpose of deceiving gullible people. Their methods are very predictable. The list below amounts to a how-to advice manual for committing logical fallacies. Before doing this, a word of caution is needed. Just because you find one of the marks of pseudoscience in an article you read or in something you hear in the mass media, it does not mean you are being exposed to pseudoscience. Indeed, most new science has at least one or two of the marks of pseudoscience. As was said in the introduction, if something sounds like a duck, it may or may not be a duck. But if it quacks like a duck, smells like a duck, walks like a duck, has feathers like a duck, and contains genetic material like that of a duck, one can assume that it is indeed a duck. Such is the case with identifying junk science. If your friend tells you about the latest miracle treatment or the most recent amazing discovery, your ears should prick up. If your skepticism antennae detect three or four of the characteristic traits of bogus scientific claims, you can be assured it is pseudoscience. Below is a list of the marks of pseudoscience, followed by a definition and some examples of each.

1. A lack of reproducible evidence
2. Overreliance on anecdotal evidence
3. Play on (supposed) inconsistencies in science
4. Attempt to explain the unexplainable/appeal to myth or mystery
5. Irrefutable hypothesis
6. Argument by analogy
7. Abuse of well-known scientists
8. Misuse of statistical arguments
9. Filtering data—the "grab-bag" approach to data
10. Refusal to revise, even when proven wrong
11. Going directly to the media rather than peer-reviewed journals
12. Claims of suppression/intimidation/persecution
13. Claiming to have discovered a new paradigm
14. Claiming to have a cure for seemingly every illness

A Lack of Reproducible Evidence

The first of our characteristic marks of pseudoscience is the one that defines the subject. It is perhaps the most difficult to discover. This is because it is more difficult to discover the *lack* of something than the presence of something. This fact is employed, sometimes quite cynically, by many purveyors of pseudoscience. For example, astrologers will claim that astrology is very scientific and supported by many studies, but then will not list any of these supposed studies. Either that, or they will perhaps make casual reference to a study that was rejected by science, but astrologers will not actually mention any of the evidence found in this study. In Hagelin's defense of the Maharishi Effect, he mentions "41 previous studies." We are left to imagine the quality of these studies, but are given no details of the data.

Overreliance on Anecdotal Evidence

An anecdote is a personal story. In the context of pseudoscience, anecdotal evidence is employed when personal testimony is used to support a particular claim. It is not as if scientists totally reject the relevance of anecdotes. Many examples could be listed, where a researcher heard a story of a particular effect and became motivated by the story to pursue the question scientifically. An interesting example of this is the discovery of the ability of dogs to detect certain types of cancer. A number of doctors reported cases of patients telling them that their dog would not stop sniffing or licking a sore on their body, only to later discover the place being sniffed or licked was a nascent cancer. These anecdotes eventually motivated a careful study of dogs' cancer-detecting abilities, which produced a positive result. It is possible a scientist will mention an anecdote in the background introduction of his or her research paper. This is what happened when the dog/cancer paper was published in the *New*

England Journal of Medicine. Anecdotes, however, are NOT accepted as scientific evidence.

The use of anecdotal evidence is one of the most common marks of pseudoscience. As a rule, all pseudoscientific "alternative" medical claims rely heavily on anecdotes. A recent report on homeopathic medicines began with the statement that "Great Britain's Royal Family, Mahatma Gandhi, John D. Rockefeller, Tina Turner and Yehudi Menuhin … all have been strong supporters of homeopathic medicine." Is this supposed to be a reason we should accept homeopathic treatments to be scientific? The author expects us to reach this conclusion. Another example of the use of anecdotal evidence is found in UFO claims. It would be fair to say the *only* evidence for UFOs is anecdotal.

Play on (Supposed) Inconsistencies in Science

One of the common ploys of those trying to soften us up for belief in a nonscientific claim is to undermine our confidence in the reliability of standard scientific claims. The scientific community is labeled as a closed community of believers who give the false impression their conclusions are more reliable than they really are. Of course, this charge is not completely without merit. Scientists have behaved this way at times, but really, this is beside the point. The question is not whether the current scientific explanation is the final answer. The important question is whether the proposed alternative is a better one. The argument goes something like this: Even scientists do not agree on the explanation of such-and-such phenomenon. Therefore, my explanation is just as reasonable as theirs, and deserves equal consideration. This is a clear logical fallacy. As a rule, when scientists do not agree on which of two alternative explanations to accept, it is because there is reasonable scientific support for both explanations. Both fit some of the data, but neither fits all the data. The problem with the pseudoscientific explanation,

generally, is that it fits *none* of the data! The claim to deserve equal consideration is completely bogus.

Often the claim of inconsistency in science involves a debate that was settled a long time ago. This is why the word "supposed" was used in this mark of pseudoscience, because the inconsistency in science is often not even a real one. This pseudoscientific trick is to use an **anachronistic argument**. An anachronism is a thing that is out of place in time. An anachronistic argument in this context is one that is being held, even though it was settled long ago. We will see an example of this below.

The tendency to soften the ground of belief by attacking our faith in science is common in medical claims. Supporters of magnet therapy and many other alternative therapies try to lay groundwork for belief in their alternative treatments by reminding us of the failures of "traditional" medicine. "Western medicine often seeks to treat the symptom, not the cause." There is truth in this criticism, but unless one has evidence from double-blind experiments that magnet therapy can actually reduce swelling from arthritis, such criticisms are not relevant. In the book *Scientific Creationism*, author Henry Morris tries to undermine our confidence in the usefulness of radiometric dating of fossils by pointing out that "It is common to find that the several ages that are obtainable from a suite of uranium-thorium-lead isotopes are either discordant among themselves or 'anomalous' with respect to the assumed age of the formation." In this case, the criticism of the traditional science is legitimate, although exaggerated. Anomalies in radiometric fossil dates are not uncommon; however, Morris wants us to believe the fossils in question are only five thousand years old. Where is his data to support this conclusion? This is a clear example of playing on the inconsistencies of science as a diversion tactic to a lack of support for a pseudoscientific claim.

Another example of playing on supposed inconsistencies in science is found in a quote from a member of the Flat Earth Society who is trying to "prove" the earth is flat and that it is the center of the heavens. "The astronomers cannot even agree among themselves. Take the question of the sun's distance from the earth. Copernicus computed it to be 3 million miles, Johannes Kepler said it was 13 million miles; Sir Isaac Newton, 54 million miles; and a more recent astronomer, 104 million miles. Copernicus claimed that the sun was stationary; the eighteenth century astronomer Sir William Herschel claimed that the solar system as a whole was moving. Says Voliva: 'It is asserted by these advocates of the Copernican system of astronomy that it is an EXACT SCIENCE—and yet these two great men, Copernicus and Herschel, contradict each other, Copernicus saying that Herschel is a liar and Herschel saying that Copernicus is a liar—and Voliva agrees that they are both right.'"[1]

Here, Voliva is using an anachronistic argument. He is trying to create the false impression that the distance from the earth to the sun is in doubt. We now know the distance to the sun to within a tiny fraction of 1 percent, yet Voliva tries to create the false impression that scientists do not know what they are talking about. Another point to be made here is that the supposed argument between Copernicus and Herschel is a false one. Herschel lived nearly three hundred years after Copernicus. We can assume that, if Copernicus had available to him the evidence at hand in the time of William Herschel, he would have agreed with all modern scientists that the sun does in fact move around our galaxy. In the end, why do we reject the argument of Voliva? Because he presents no evidence for his conclusion.

Another common example of playing on supposed inconsistencies in science is in purveyors of supposed "alternative" medicines. First of all, we might want to ask, "What is 'alternative' medicine?" Is there such a thing as "alternative" physics or chemistry or biology? "Alternative medicine" is, generally, a red flag for unscientific medicine. In any case, a common

practice of the supporters of megadoses of vitamins, homeopathic medicine, and the like is to undermine our faith in "Western" (i.e., scientific) medicine. They will point out (perhaps with some validity) that "Western" medicine goes after the symptoms, rather than the cause. They will point out that pharmaceuticals often have terrible side effects that, at times, cause more harm in the long term than any benefit. All this is true. Here is the point. How is this evidence that we should seek chiropractic, homeopathic, or other "alternative" medical treatment? This is "playing on the supposed inconsistencies in science."

While we are on the subject, let it be said that when you hear the claim "this medicine has no side effects," you should assume it also has no effects. Any physiologically active drug (in other words, any medicine that has any effect at all) will have side effects.

Attempt to Explain the Unexplainable or Appeal to Myth or Mystery

Scientists have been able to explain many of the great mysteries, such as why the planets move in elliptical orbits, how babies are produced, why infectious diseases spread, and why there are volcanoes. Nevertheless, there remain mysteries to be solved by science. How do we remember? What is the nature of human consciousness? In addition, history provides us with many mysteries that scientists may never solve. Why did the ancient Mayans produce half-bird, half-reptile carvings on their temples? How did the Greeks create such interesting computer-like machines? Add to this an almost limitless variety of myths of creation, cataclysmic events, unicorns, and other monsters. The response of good science to such mysteries is to provide speculative theories (and to confess the theory is highly speculative), or to simply refrain from offering a theory at all, given the lack of sufficient data.

This is not the response of those susceptible to producing pseudoscientific explanations. For them,

---

1 *Leaves of Healing*, May 10, 1930, p. 135.

mysteries such as the manna falling in the desert in the biblical book of Exodus are a feeding frenzy for scientific-sounding explanations. Why is it that the Babylonians began to worship the planet Venus in the second millennium BC? Science has no explanation, but Immanuel Velikovsky, in his book *Worlds in Collision*, offered a good explanation. Venus was a giant footloose comet that nearly collided with Earth a number of times in the fairly recent past. One of its near-collisions resulted in fervent worship of the planet by the Babylonians. Another near-collision around 1400 BC can explain the falling of the manna to Earth. This was due to the condensation of organic matter from the tail of the comet, which became Venus! Velikovsky began with his planetary collision scenario and proceeded to look at cultures around the world, seeing if he could explain otherwise mysterious historical facts. Having gotten started, he claimed near-collisions for Mars. This can explain why Mars, which once was wet and had an atmosphere, is now much colder and drier. Earth captured its atmosphere in a near-collision.

How will we explain the half-reptile, half-bird relief carvings on a Mayan temple in Totonacapan in present-day Veracruz, Mexico? Simple: it is the representation of a dinosaur-like creature that lived a few thousand years ago in the jungles of Mexico. This is the explanation offered by Henry Morris in his book *Scientific Creationism*. One can only suspect this explanation came not from the evidence, but from Morris's desire to undermine the belief that dinosaurs went extinct sixty-five million years ago.

Then there is the mystery of the fantastic "fine-tuning" of the universe. The gravitational force, the strong and weak nuclear forces, the nearly perfect balance of matter and antimatter, and more than a dozen other parameters appear to be precisely tuned. If any of them were changed only slightly, it would result in a universe without stars, galaxies, and obviously without living things to observe that universe. The questionable explanation offered to this mystery

is the multiverse theory. There are an infinite number of universes, and we happen to live in the one with the right properties. Is there any evidence for these parallel universes? The answer is no. Might there be a more reasonable explanation for the Mayan carvings? Is there a believable model for how organic material in a comet tail could produce edible food? The answer to all of these is no, but this does not bother those who produce such theories.

A fun example of "explaining the unexplainable" is the case of the Sirius Mystery. A fellow of the Royal Astronomical Society (London), Robert K. G. Temple, informs us of an interesting and mysterious discovery from the Dogon tribe in Mali, West Africa. The Dogon have a myth about a star they call *Digitaria* or *po tolo*. It is the details of the myth that make it a good opportunity for pseudoscientific explanation. The Dogon believe *Digitaria* is a very small, invisible star that circles the well-known star Sirius. Sirius is one of the brightest stars in the Northern sky. According to the Dogon, *Digitaria* is "the smallest thing there is" and yet also the heaviest star. They believe it revolves around Sirius once every fifty years and that its orbit is not circular. It just so happens that Sirius is a binary star. The twin star to Sirius is called Sirius B, which is not visible to the naked eye. Being a white dwarf, it is relatively very small and has a density more than a billion times as great as our sun. The orbit of Sirius B is elliptical, not circular, and its period of orbit is roughly fifty years. How are we to explain this amazing coincidence? The Dogon believe there is a third star in the system, as well as a planet called Nommo, whose inhabitants have visited the earth. Temple suggests that this cannot be ruled out, as it is the only "reasonable" explanation of the facts. To the pseudoscientist such a mystery is an opportunity to spin an interesting (and possibly lucrative) explanation. To the scientist, it is an interesting story, but Temple's explanation is definitely not scientific and, in the absence of any evidence of actual visitors for Sirius B, they will wisely defer providing a scientific-sounding explanation.

In each of these examples, wise scientists will refrain from speculative explanations or ones that just happen to support their pet theory. When we come across speculative explanations of unsolved mysteries and myths we have heard the sound of a duck-like quack. Is this alone proof that the proposed theory is pseudoscience? Maybe not, but it should give us pause to be suspicious.

Irrefutable Hypothesis

Philosopher of science Karl Popper proposed that the demarcation between science and nonscience is falsifiability. If a theory is not refutable by any conceivable experiment, then that theory is—by definition—not a scientific one. If a theory is, in principle, refutable, but cannot be falsified by any experiment with which scientists have the technological ability to perform to date, then that theory is at best speculative. The irrefutable hypothesis is another of the favorite tools of pseudoscientists. It is often found in concert with the use of myth or mystery.

Immanuel Velikovsky insists Venus was a comet that nearly collided with Earth in the recent past. Can scientists offer a single experiment to disprove this theory? The UFO believer challenges science to produce a single experiment disproving the reality of alien visits. Ancient astronaut theorists such as Hal Lindsey insist science has yet to prove their ideas to be false. The universe we live in may very well be the result of intelligent design, but is the theory of intelligent design a scientific one? Probably not, because design cannot be refuted by any conceivable experiment. This theory becomes pseudoscience when its supporters propose that it is a scientific explanation, rather than a metaphysical one.

Argument by Analogy

Argument by analogy is one of the more subtle indicators of pseudoscience, but it is also one of the more common. The argument goes something like this: The cause-and-effect relationship I am proposing is almost an exact analogy to a well-documented and widely accepted scientific explanation of another cause-and-effect relationship in nature. Therefore, my theory is very likely a good explanation of the effect in question—it is scientific. This is another example of a logical fallacy. The mere fact that two explanations are logically analogous and that one explanation is known to be valid does absolutely nothing to prove the *other* explanation is valid. In science, an explanation is considered valid if it is supported by well-documented, reproducible scientific evidence.

How can we be assured that homeopathic remedies really are effective? Dana Ullman, MPH, assures us that "Conventional medicine uses homeopathic-like therapy in choosing radiation to treat people with cancer (radiation causes cancer), digitalis for heart conditions (digitalis creates heart conditions) and Ritalin for hyperactive children (Ritalin is an amphetamine-like drug which normally causes hyperactivity)." Never mind that these analogies to the homeopathic "law of similars" are dubious. Even if we can concede the analogy is legitimate, this provides no evidence whatsoever that bee venom can cure a sore throat. Only a double-blind study can establish this treatment as scientific. Ullman offers no such evidence.

In the 1980s and '90s many otherwise intelligent and rational people came to believe in biorhythms. Biorhythms were popularized in the United States by the book *Is This Your Day?* by George Thommen, published in 1964. According to the biorhythmists, there are three great "rhythms of life." There is a 23-day physical cycle. From the day of our birth, we experience a periodic increase and decrease in feeling energetic, says Thommen. Then there is the 28-day emotional cycle, as well as the 33-day intellectual cycle. Everyone has their good and their bad days, physically, mentally, and emotionally, and the three usually do not correspond. Biorhythms can explain this fact. Do biorhythmists have double-blind studies that support their claims? None have been published so far. What they do have as "evidence" to support belief in biorhythms is the fact that scientists have discovered many periodic effects among both plants and animals. Such phenomena have been studied extensively and the means of telling time by living things has been studied. In fact, there is an entire field of "chronobiology" that studies circadian rhythms (with a period of one day), infradian rhythms (a period of less than one day), and ultradian rhythms (a period of more than one day). We all know our serotonin and melatonin levels go up and down, governing our sleep cycle.

What does this have to do with biorhythms? Nothing. But if you read the literature of biorhythmists, they imply that biorhythms are one of the examples of ultradian rhythms. They imply that the fact cicadas come out every seventeen years is evidence, by analogy, to support biorhythms. In the absence of reproducible, scientific evidence, such argument by analogy is a logical fallacy and is definitely not scientific.

Abuse of Well-Known Scientists

One of the red flags indicating unsubstantiated, unscientific claims is the spurious use of well-known scientists as a sort of character reference to such claims. We are all quite familiar with mass media advertisements using respected sports or other entertainment figures to hawk their wares: if Viagra worked for Bob Dole, it will surely work for you.

There are several problems with the use of such arguments in science. First, there is the fact that science does not acknowledge authority. The reputation of the name behind a particular theory does not lend greater weight to the scientific evidence supporting that theory. Then there is the questionable nature of such name-dropping as it is practiced in typical pseudoscientific literature. Not yet convinced that magnet therapy will solve your lower back pain problem? It might help to know that the same theory Albert Einstein used to unlock the power of the laser is used to explain the miraculous healing power of magnetic fields. Such unscientific arguments take two forms:

1. Quoting a scientist out of context.
2. Name-dropping: mentioning a famous scientist in connection with a theory when that scientist would certainly not authorize his name being used in connection with the claim.

Henry Morris quotes a leading expert on overthrust faults who tells us that such faults seem at first glance to undermine the entire notion of index fossils. The aim of Dr. Morris is to undermine our confidence in the accepted view of the age of the earth. At the end of the quote we are left wondering what is the next word immediately following the quoted material. Our suspicion is confirmed when we go to the actual reference; the word following the quotation is "but." The expert on overthrust faults has been made to say the opposite of what he intended by the use of selective quotation. This kind of irresponsible behavior is common in pseudoscientific writing.

Misuse of Statistical Arguments

It is normative that scientific claims about the validity of connection between independent and dependent variables must be analyzed statistically. The typical consumer, however, is not aware of the subtleties of statistics. For the one trying to create a case for unscientific claims, the bogus use of statistics is a powerful tool. John Hagelin and his research team report the rate of violent crime decreased by 18 percent in the Washington, D.C., area when TM-Sidhi was practiced. Is this evidence that TM-Sidhi was the cause of this drop? Several questions come to mind. Are there any hidden variables—other causes in operation that might have produced the observed effect? Was there a control group? If not, why not? If there is no control group, then it is reasonable to ask this question: "The crime rate was 18 percent less than what?"

The literature at your local GNC claims a certain herbal medicine increases memory retention by more than 30 percent. You would be wise to be skeptical of such a statement at face value. No herbal remedy has been proven to prevent cancer. This is not a problem to the pseudoscientist. He can tell us the plant has been used by a remote tribe for decades, and that this tribe has a statistically proven low historical rate of cancer. Darrell Huff wrote a classical introduction to statistical treatment of data with the title *How to Lie with Statistics*. Huff instructs his readers on how to make a statistical correlation appear to be a cause-and-effect relationship. A search on the Internet for the spurious claims of alternative medical therapies gives one the impression the authors of these claims have been carefully studying Huff's book. Are these people charlatans, lying in order to make a fast buck? Are they sincere but self-deceived believers? Or alternatively, are they promoting something genuine that is not yet supported by sufficient good data? In any case, let the buyer beware.

An example of questionable statistics being used to support a pseudoscientific claim comes from parapsychology. Claimed phenomena in this category include telepathy (reading the thoughts of others), clairvoyance (knowledge of the future), and telekinesis (ability to use mental power to move physical objects). Yuri Geller was a famous practitioner of telekinesis and other "psi" abilities. He gave traveling shows of his power as a mentalist. Of all the areas that fall under the rubric of pseudoscience, perhaps parapsychology has engendered the most attempts at actual tests of its

veracity. In fact, until the 1980s there was a research group under Dr. J. B. Rhine studying parapsychological research at Duke University. Practitioners of this "embryo science," or "frontier science of the mind," have their own journals, such as *Journal of Parapsychology*, *Journal of Near-Death Studies*, *Journal of Consciousness Studies*, and *Journal of the Society for Psychical Research*.

The reason this is interesting to us is because a number of studies have been published by Rhine and others that, at first glance, appear to support the conclusion that some humans have telepathic or clairvoyant abilities. Studies have been done using a "sender" in one room and a "receiver" in another. Rhine associate J. G. Pratt used cards with five different symbols. The "receiver" was in a different room. The experiments involved having the subjects sit at varying distances, on the hypothesis that the rate of success would decrease with distance. Some studies gave remarkable results with the accuracy of the guesses having odds of millions to one. The problem with such studies is they tend to not be reducible. If the same "sender" and "receiver" are tested repeatedly, eventually the rate of correct guesses approaches randomness. In the end, we are left with results that do not hold up to rigorous statistical analysis. In other words, the results are not scientifically verifiable.

Filtering Data—the Grab-Bag Approach to Data

By now you probably have the impression that many of those making unscientific claims can be accused of playing fast and loose with the evidence. This is not always the case, but there is a noticeable pattern. Many pseudoscientists tend to notice only such evidence that supports their beliefs and conveniently filter out the evidence that does not support their preconceived conclusions. In his book *The Borderlands of Science*, Michael Shermer calls this a "knowledge filter." We are going to call this the grab-bag approach to data.

Believe it or not, many who do this are not being willfully deceitful. It is their methodology—not their honesty—that is faulty. Science is a search for consistency. We already know that legitimate scientists do not always behave ideally, but philosophically they hold to the idea that they pursue the data wherever it leads them. The problem with many purveyors of pseudoscience is that they are in search of proof, not in search of truth. Their goal in looking at the body of evidence relevant to a particular question is to find possible ammunition to prove that what they believe is for reasons other than the data. It is not the goal here to undermine the importance of faith. Not at all. In fact, we have already established that scientific pursuit itself is based on a kind of faith in the fundamental presuppositions of science. In scientific problem solving, however, such faith needs to be left to the side.

Our first case study is a classic example of the grab-bag approach to data. The work of Hagelin *et al.* can be described as pathological science. These researchers appear to have a religious commitment to believe in the efficacy of TM toward improving humanity. Experience with such true believers is that they are impervious to data contrary to their hoped-for conclusion. Therefore, when they found the homicide rate increased dramatically over their test period, they naturally found themselves explaining this away. What would they have done if their time-averaged statistical analysis had shown an increase in numerous other crime statistics as well? Would they have published their results and called the press conference? Would they have kept working over the data to create a better spin?

It has been the official policy of the *Journal of Parapsychology* to reject for publication papers with a null or negative result. Experiments that do not support belief in psi are not considered evidence that psi is not real. Instead, they are labeled as "failures." This is a rather blatant example of the filtering of data. Young earth creationists tell us that radiometric dating methods are sometimes unreliable. The implication

is that they are often reliable. The fact is that when parallel methods are used, they are usually quite reliable. What do the young earth creationists do with this reliable data? They find an excuse to ignore it.

Young earth creationist and geologist Dr. Henry Morris published data on the content of certain minerals in the ocean, arguing the mineral content in the oceans supports the conclusion they are only a few thousand years old. He divides the current total content of an elemental mineral such as potassium, sodium, iron, or nickel in the oceans by the amount known to enter the oceans per year. If one divides the total amount of nickel, for example, by the amount of nickel entering per year, one can calculate the number of years nickel has been accumulating in the oceans. The data supports an age for the oceans, based on the nickel, lead, magnesium, and copper content of a few thousand to a few tens of thousands of years. Does this support a conclusion the oceans are only a few thousand years old, as Morris suggests? Maybe not. The fact is that the ions listed above are extremely insoluble in carbonate solution and the ocean is loaded with carbonate ion. Any geologist knows that these ions could not build up in the oceans, no matter how old they are. Add to this the fact that the content of more soluble ions such as lithium, sodium, potassium, and chloride give a calculation for the age of the oceans of tens or even hundreds of millions of years. On balance, the ion content of the oceans strongly supports the conclusion they are at least a few hundred million years old. Why would a scientist ignore the data implying a great age and present the anomalous data (which is, in fact, not anomalous)? This is the behavior we are labeling the grab-bag approach to data.

Refusal to Revise, Even When Proven Wrong

The work of those who make it their aim to debunk unscientific claims can be very frustrating. One of the cardinal tenets of science holds that faulty results and explanations must be retracted or revised. As a rule, pseudoscientists do not play by this rule. When their experiments are found to be unreliable or when their speculative explanations are revealed to be full of holes, they rarely publish retractions. Even if they do, they certainly do not change their overall view of the matter.

Homeopathic medicine provides a good example of this principle. Samuel Hahnemann, the founder of homeopathy, first published the "law of similars" in 1810. By this principle, a disease with a particular symptom is cured by giving the body an extremely low dosage of a poison or irritant, which causes a similar symptom. What is difficult to swallow is the claim of homeopaths that the lower the dosage, the stronger the curative effect. For example, to cure the flu a patient may be given a 30X dilution of duck liver. A 30X dilution is a sample that something has been diluted by a factor of ten 30 consecutive times. A major problem arises with such a dilution and the claims of its efficacy. If one were to take a pure sample of duck liver and perform a 30X dilution, the result is that even for the highest concentration substance in the original sample, the concentration in the 30X dilution would have much less than a single molecule in the sample. In other words, these homeopathic treatments are placebos. We all are well aware of the power of placebo treatment, but homeopathic theorists seem to ignore this factor.

When Hahnemann proposed the law of similars in 1810, chemistry was at such a primitive stage that the number of molecules in a sample could not be counted. Perhaps we can forgive Hahnemann for his lack of knowledge in this area. Nevertheless, one hopes modern homeopaths would abandon their theories once it is shown beyond a doubt that their treatments are diluted to the point where the treatment concentration is literally zero. One would expect them to revise when proven wrong. Unfortunately, this is not the case. Rather than abandon their disproved

theory, they have proposed that the water retains the memory of the original substance. To quote Dana Ullman: "Although the homeopathic medicines may be so dilute as to not have any molecules, a pattern of the substance remains." I will let this statement speak for itself.

Going Directly to the Media Rather than Peer-Reviewed Journals

Pseudoscientists complain the scientific establishment has shut them out of mainstream journals. This criticism is well founded. Pseudoscientists are caught between a rock and a hard place, as they are criticized for not publishing in peer-reviewed journals. Yet those in charge of these journals refuse to publish research in areas having the aura of pseudoscience. Of course, the net result is to keep a lot of very bad science out of the literature, but what if a claim labeled as pseudoscience is, in fact, legitimate? Whether this treatment is justified is something we can debate. Nevertheless, one of the consistent characteristics separating legitimate science from pseudoscience is found in analyzing where the research is published. The Washington, D.C., TM experiment is a case in point. When researchers call a press conference, the scientist's skepticism detection kit goes on high alert. Consider the announcement of the first cloned human by the shadowy group known as the Raelians. On December 27, 2002, Brigitte Boisselier, a chemist for the Raelian-sponsored company Clonaid, announced the birth of the first cloned human, named Eve, the previous day. The actual name and nationality of the supposed human clone was not revealed. We are still waiting.

By this time, you probably get the point. For the last few of our "marks of pseudoscience," I will only give a brief definition and will not provide case studies.

Claims of Suppression

To deflect attention away from a claim with very weak scientific support, the author or spokesperson will complain of being suppressed or persecuted by the scientific establishment. The claim of intimidation from the scientific community may or may not be legitimate, but in any case this does nothing to establish the reliability of the claim.

Claims of Discovering a New Paradigm

One of the most prevalent strategies of pseudoscientists is to claim to have found a principle that is completely new to science. The declaration of such scientific revolutions are rarely proved true over time. We know from our history of science that brand new scientific ideas are often resisted strongly by the scientific establishment. The difference between a scientific genius and a crank are not always as great as we would like, but beware of those who claim to have a brand new idea or a new paradigm. The claim of suppression and of having a brand new paradigm often come hand in hand.

Claiming to Have a Cure for Virtually Every Illness

Homeopathy, chiropractic, acupuncture, and other alternative medical sciences have one thing in common: if their claims are to be believed, these therapies can cure pretty much everything. They offer the solution to arthritis, senility, gout, chronic headaches, sinus infections, diabetes, lupus, and even heart disease and cancer. Here is a good rule of thumb (although not a scientific statement): those who claim they can cure everything can probably cure nothing.

Only True Believers in the Research Groups and Lack of Scientific Credentials

Our last two marks of pseudoscience are related. The fact that a researcher is paid by the Magnet Therapy Institute does not prove the research is bogus, but it is at least noteworthy. If you do an Internet search, you often find the credentials of those who promote the latest fad diet or the "new physics" of pyramid power are not in nutrition or in physics—or if they are, the degree is not from a credentialed university. Lack of credentials does not prove claims are unscientific, and the possession of legitimate scientific credentials does not create immunity to pathological science. John Hagelin has absolutely the best scientific credentials. It is wise for us, however, to consider the source of a too-good-to-be-true claim.

Borderline Cases of Pseudoscience

We have devoted considerable time warning against unsupported claims outside the mainstream of science. You have been trained to be very skeptical of such claims, and for good reason. Let us finish our treatment of pseudoscience by opening the door just a little. We will now offer the possibility that some claims appearing to be pseudoscience might be "borderline cases."

First, let us propose possible parameters for establishing something that, by definition, is not now scientific, but may just possibly turn out to be legitimate.

1. An apparent phenomenon with no known cause/mechanism.
2. An apparent phenomenon that, if it were true, would require a new scientific paradigm.

One person's "borderline case" might be another person's outlandish proposal. The reader will make his or her own choices here. But we have precedent to consider. Our case study will be continental drift, also known as plate tectonics. When first proposed by Alfred Wegener, this theory had many of the marks of pseudoscience. In fact, it was pseudoscience by our working definition. As we will see, it was an "apparent phenomenon" in that there was sufficient information to force many to consider the idea, but the motion of landmasses had no known cause. In fact, it had no conceivable cause at the time. The continental drift theory fit the second parameter as well. If it were true, it would have required a new scientific paradigm. Now, nearly a hundred years after Wegener's book was published, it has indeed become one of the two paradigms of geology.

We will get back to our case study, but first I will propose some possible borderline cases for consideration. Again, one person's borderline case is another person's nonsense idea. You, the reader, will have your own list.

Our first proposed "borderline case" is herbal medicines. Probably most of these untried and unproved supplements have no reproducible effect. Let us concede that a good majority of "traditional medicines," if studied under well-controlled, double-blind studies, would fail dramatically (as do most pharmaceutical trials of treatments devised by traditional medical research, by the way). After conceding this point, experience informs us that almost certainly some of these herbal medicines will indeed have some effect. Aspirin is synthesized from salicylic acid, which is found in willow bark. Willow bark was a herbal remedy for centuries before it became a scientific treatment for pain. Another example is the natural alkaloid quinine. The active ingredient in quinine water is found in the bark of the cinchona tree. It is a natural anti-inflammatory, analgesic, and most important, an antimalarial drug. It was used for hundreds of years by native groups. By definition, its

use was pseudoscientific, in that it was not tested in scientific studies until relatively recently.

Any untested herbal remedy claim is, by definition, pseudoscientific. Perhaps we should be more cautious, and call such treatments "not yet scientifically verified." Only in the last twenty years or so have mainstream scientific organizations such as the NIH begun to study herbal medicines systematically. Let's be honest: there is considerable reluctance for mainstream science to study these products, because they have achieved the dubious label of pseudoscience. Some herbs, such as echinacea, a proposed treatment for the common cold, have not held up well to scientific scrutiny. Our first proposed borderline case fits the first parameter above. Until a specific compound in the proposed herbal remedy (such as quinine) is isolated and studied for its effects, these remedies have no known cause. Most herbal remedies are "apparent phenomena with no known cause."

Now we will step out on a bit of a longer limb and consider two other possible borderline cases, which fit both the first and second of our proposed parameters: chiropractic and acupuncture. If the broader claims of either of these two therapeutic approaches are true, they will require a scientific revolution and produce a new medical paradigm.

Both of these alternative therapeutic approaches have achieved a certain level of acceptance by the medical establishment. In some cases, they are even covered by medical insurance. We will not delve into the details here, but it seems unreasonable to completely deny all efficacy to these therapies. Surely we can concede that chiropractors can achieve success in dealing with lower back pain. Perhaps critics will propose the only thing acupuncturists produce is a very strong placebo effect, but given the fact that acupuncture is occasionally used in lieu of anesthesia in surgery, this seems to be a stretch. That would be a very powerful placebo effect.

Let us, for the moment, accept that both chiropractic and acupuncture have achieved sufficient success to rise to the level of being "apparent phenomena." If this is true, what is the mechanism by which they achieve these effects? True believers in chiropractic or acupuncture claim they can prevent or cure virtually every illness with these treatments. They do so with essentially no side effects. This should make us be a bit skeptical. The great majority of their claims have not been verified by experiment. This is even greater reason for skepticism. At this point, most (but arguably not all) of the claims of these two therapies are pseudoscience. Both chiropractic and acupuncture theorists propose mechanisms involving certain types of energy that have never been demonstrated to exist. Acupuncturists claim to control the flow of "chi." What is chi? What is its color? What is its density? Where are the channels through which it flows? Is there any evidence that chi even exists? What is this evidence?

In the final analysis, if either chiropractic or acupuncture ever achieve full status as scientific models, they will require the acceptance of a new paradigm. Who knows? Maybe chi theory will be the third medical paradigm, after germ theory and genetic disease. Good science is open to this possibility.

A Case Study: Continental Drift

In 1914 Alfred Wegener, a young meteorologist, was hospitalized for several months for major injuries he sustained as a soldier in the German army. While recuperating, he wrote and eventually published (in 1915) a book titled *The Origin of Continents and Oceans*. He proposed that Africa and South America formed a single landmass at a point in the very distant past, and that these two continents had moved away from each other very slowly over vast periods of time. His was the first serious proposal of what we now know of as the theory of continental drift. Was Wegener a crank or a revolutionary? Some in the scientific community received his claims with skepticism, but also with an open mind. The majority, however, especially those in geology, called his proposal ludicrous. They attacked Wegener's methodology as unscientific. "Utter damned rot," said the president of the American Philosophical Society. "If we are to believe [this] hypothesis we must forget everything we have learned in the last 70 years," said an American scientist. According to a leading British geologist, no one who "valued his reputation for scientific sanity" would support such a theory. In the end, Wegener's theory was accepted, but we will see that those who called him a pseudoscientist did not do so without cause. His work exhibited some of the marks of pseudoscience.

Even if his proposal was not strictly scientific at the time, Wegener had sufficient evidence for his proposal to deserve to be labeled as a "borderline case." His was an "apparent phenomenon with no known cause." Wegener's evidence included the following:

1. The apparent coincidence of the shape of western Africa and eastern South America.
2. Fossil evidence for species such as the plant *glossopteris* and the marine reptile *mesosaurus*, which were known only in parallel regions in South America and Africa.
3. Living species such as lemurs, which only live in Madagascar, but whose fossils are also found in relatively recent formations in western India.
4. Evidence for horizontal forces and motion (at least on a local level), such as overthrust faults.

After his book was published, but before his death, Wegener added geoglacial evidence of parallel ice ages to his data.

Were his critics being overly critical and closed-minded? Perhaps, but their skepticism was not completely without reason. Consider some "marks of pseudoscience" in Wegener's proposals. First of all, his work had at least the appearance of using the grab-bag approach to data. Of the four lines of evidence above, one is from geography, one from paleontology, one from botany, and one from geology. Such interdisciplinary "research" was not normal in his day. In this sense, Wegener was ahead of his time. To his critics he appeared to be following the familiar pattern of pseudoscientists.

Another mark of pseudoscience that Wegener was accused of is making scientific claims in an area where he had no expertise. What is a meteorologist doing publishing geology theories? With hindsight, we easily forgive Wegener of this indiscretion, but geologists at the time were suspicious.

Then there is the fact that his theory was irrefutable. From the point of view of his critics, it was very convenient—bordering on suspicious—that his explanation was not testable. The distance between the continents was not known with sufficient precision to measure the proposed extremely slow movement.

In Wegener, we have another mark of pseudoscience. He was persecuted, hounded, criticized, and blocked from the inner circles of scientists for the rest of his life.

The greatest criticism of Wegener's proposal was that he had no believable mechanism by which entire portions of the earth's crust could move thousands of miles. Wegener proposed that perhaps the tides could move the continents. Alternatively, he proposed that the continents moved away from each other from an initial location at the North Pole, due to centrifugal forces from the spinning of the earth. Wegener called this *Pohlflucht* (flight from the poles). Critics scoffed at these proposals, and physicists showed the forces produced could not even come close to creating the motion. In the end, Wegener was forced to admit he had no mechanism to propose to explain continental drift. In the last edition of his book, he conceded, "It is probable the complete solution of the problem of the forces will be a long time coming." His proposal was "an apparent phenomenon which, if it were true, would require a new scientific paradigm." In fact, this is exactly what eventually occurred.

Why is continental drift, otherwise known as plate tectonics, now accepted as one of the two paradigms of geology? Is it because Wegener wore down his critics, or that they gradually became more open-minded? No. The ultimate victory for his theory was achieved because eventually more anomalies arose, which the static earth theory could not explain, and greater evidence arose that was consistent with Wegener's hypothesis. This is the normal process by which science evolves. In the 1950s the Mid-Atlantic Ridge was discovered. One striking piece of evidence in support of plate tectonics is the fact that the rocks in the ocean floor near the ridge are quite young. They get progressively older symmetrically, going east or west from the ridge, all the way to Africa and South America. Evidence from geomagnetism, ice age location, motion of the magnetic North Pole, and many other kinds of evidence eventually poured in. By about 1960 continental drift was part of scientific orthodoxy. In the 1980s NASA placed a mirror on the moon. Scientists can now use lasers bounced off this mirror to measure the distance from Africa to South America with such amazing precision that they can actually measure the rate at which they are moving apart.

Was Wegener a crank and a pseudoscientist, or was he a revolutionary? History has granted him the latter label. What is ironic is that Wegener was a scientific revolutionary, yet his initial work was not fully scientific, at least in a narrow sense. It is a sad final note to this story that Wegener died in 1938 on an expedition to Greenland to further test his theory. Others in his expedition got separated from the rest of the party. Wegener died in a heroic attempt to save his colleagues.

What can we conclude from this case study? First, Wegener was not a pseudoscientist, no matter what his critics said. In the bigger sense, we should be reminded that just because something is not scientific does not mean, *a priori*, that it is not true. Certainly, nearly all of the pseudoscientific theories out there are sheer nonsense, and the majority of their supporters are pseudoscientists. But in the final analysis, even in our criticism and our well-justified skepticism of these unscientific claims, a measure of humility and open-mindedness must remain part of the scientific approach to asking and answering questions.

Section VI
Science and Religion

Introduction

By John Oakes

A lot of heat has been generated in what can be viewed as a modern battle between science and religion. Clearly, when the two are discussed in the same breath, important issues are raised. Hopefully, we can shed light on the subject, rather than generating more heat. By way of introduction, let us consider four different definitions.

agnosticism: The agnostic says we cannot know for certain if there is a God or other supernatural reality.

atheism: The atheist says there is no God or metaphysical/supernatural reality.

deism: The deist sees design in the world, and therefore, a designer. However, this creator/designer does not intervene supernaturally in the physical world.

theism: The theist, like the deist, sees design and a designer. She or he believes that this being is personal and intervenes, supernaturally or otherwise, in the world.

Atheists and agnostics make up a minority, but a rather vocal minority of scientists. The scientist and philosopher Thomas Huxley, a good friend of Charles Darwin, coined the word agnosticism. Some atheists and agnostics see religion as an intrusion, and for some an unwelcome intrusion that can impede the progress of science. You will read an essay by Norman and Lucia Hall expressing this view. As you do, ask yourself why the Halls see the religious mindset as inherently opposed to science. A fairly small proportion of humans as a whole are deists. However, many who have this perspective come to this point because of their view of science. Important figures who have been labeled as deists include Isaac Newton, Joseph Priestley, Charles Darwin, and Albert Einstein. You will be reading an essay by Albert Einstein. As you do, ask yourself what he sees the relationship between science and religion to be, and why he views the two as complementary, rather than as adversarial. Last, you will be reading an essay by a theist who sees science and religion as natural friends, not enemies. Ask yourself why he takes this position.

Is the War Between Science and Religion Over?

By Norman F. Hall and Lucia K.B. Hall

The CBS television news report, 'For Our Times," which covered a two-week conference on "Faith, Science, and the Future" held at the Massachusetts Institute of Technology, left the viewer with the feeling that the long conflict between science and religion is at an end. Hundreds of scientists and theologians gathered to discuss issues of science and ethics and proceeded from the assumption that science and religion were two non-conflicting bodies of knowledge, equally valuable complementary paths leading toward an ultimate understanding of the world and our place in it. The conflicts of the past were said to be due to excessive zeal and misunderstanding on both sides. Peaceful coexistence and even a measure of syncretism are now assumed to be possible as long as each concedes to the other's authority in their separate worlds of knowledge: that of matter and facts for science, and that of the spirit and values for religion.

Nonsense

Let us be blunt. While it may appear open-minded, modest, and comforting to many, this conciliatory view is nonsense. Science and religion are diametrically opposed at their deepest philosophical levels. And, because the two worldviews make claims to the same intellectual territory—that of the origin of the universe and humankind's relationship to it—conflict is inevitable.

Transcending the Material World

It is possible, of course, to define a non-supernatural "religious" worldview that is not in conflict with science. But in all of its traditional Western forms, the *supernatural* religious worldview makes the assumption that the universe and its inhabitants have been designed and created—and, in many cases, are guided—by "forces" or beings which transcend the material world. The material world is postulated to reflect a mysterious plan originating in these forces or beings, a plan which is knowable by humans only to the extent that it has been revealed to an exclusive few. Criticizing or questioning any part of this plan is strongly discouraged, especially where it touches on questions of morals or ethics.

Science, on the other hand, assumes that there are no transcendent, immaterial forces and that all forces which do exist within the universe behave in an ultimately objective or random fashion. The nature of these forces, and all other scientific knowledge, is revealed only through human effort in a dynamic process of inquiry. The universe as a whole is assumed to be neutral to human concerns and to be open to any and all questions, even those concerning human ethical relationships. Such a universe does not come to us with easy answers; we must come to it and be prepared to work hard.

In order to understand how scientific observations are made, let's follow a hypothetical-scientist into his or her laboratory. Suppose this scientist's task is to measure the amount of protein in a biological fluid—a common procedure in research laboratories, hospitals, and school science classes. The scientist will proceed by carefully measuring out into test tubes both several known volumes of the fluid and also several different volumes of a "standard" solution he or she has prepared by dissolving a weighed quantity of pure protein. The scientist will add water to bring all the tubes to the same volume and then add a reagent which reacts with protein to produce a blue color. After the solutions in all the test tubes have reacted for a specified period of time, the scientist will measure the intensity of the blue color with a spectrophotometer. By comparing the color intensity of the unknown solutions with that of the known standard protein solutions, he or she will be able to calculate how much standard protein is needed to produce the same color reaction as the unknown, and this, the scientist will conclude, is the amount of protein in the unknown sample.

Natural Adversaries

Now that science has won immense prestige and has become a major factor of social progress, the experts of religion are inclined to forget the forms of damning it that were used in the past and agree to all kinds of compromises. But no matter what forms the relations between science and religion take, it is clear to any sober-minded person that religion has always struggled and continues to struggle against science as its natural adversary.

—Yuri Pishchik, *American Atheist,* March 1987.

What our hypothetical scientist has done is to perform a controlled experiment. He or she must report it honestly and completely including a description or a reference to the method. He or she must also be prepared to say that all variables which could have affected the reported result, to the best of his or her knowledge and belief, have been kept constant (for example, by using a water bath to maintain a constant temperature) or have been measured (as were the different volumes of the unknown solution and standard solution) or are random (measurement errors or perhaps proteinaceous dust motes from the surrounding air). This is the essence of the scientific method.

Clearly, such a controlled experiment would be impossible if our scientist were required to entertain the possibility that some factor exists that can affect the color in the test tubes but which can never be controlled in these ways—a factor that cannot be held constant, cannot be measured by any physical means, and cannot be said to act randomly. But that is exactly what the religious, supernaturalist worldview *does* require. Untestable, unmeasurable, and nonrandom occurrences are commonplace in all supernatural religions and pseudosciences.

Fundamental Incompatibilities

This fundamental incompatibility between the supernaturalism of traditional religion and the experimental method of science has been, nevertheless, remarkably easy to dismiss. The *findings* of science over the past three centuries have been eagerly welcomed for their practical value. The *method*, however, has been treated with suspicion, even scorn. It has been perceived as being responsible for revealing the material workings of ever more of the mysteries of life which used to inspire religious awe. From the point of view of the religious believer, it has seemed as though the goal of science has been to push belief in the supernatural to ever more remote redoubts until it might disappear entirely.

This is not, and cannot be, the goal of science. Rather, a non-mysterious, understandable, material universe is the basic *assumption* behind all of science. Scientists do not chart their progress with ghost-busting in mind. Naturalism, or materialist monism, is not so much the product of scientific research as it is its starting point. In order for science to work, scientists must assume that the universe they are investigating is playing fair, that it is not capable of conscious deceit, that is does not play favorites, that miracles do not happen, and that there is no arcane or spiritual knowledge open only to a few. Only by making the assumption of materialist monism will the scientist be able to *trust* the universe, to assume that although its workings are blind and random it is for this very reason that they can be depended upon, and that what is learned in science can, to some degree, be depended upon to reflect reality.

The Unifying Theory

As evolution is the unifying theory for biology, so naturalism is the unifying theory for all of science. In his book, *Chance and Necessity*, biochemist Jacques Monod called this basic assumption "the postulate of objectivity," since it assumes that the universe as a whole is dispassionate of, indifferent to, and unswayed by human concerns and beliefs about its nature. Its inverse—in which the universe is passionately involved in, partial to, and swayed by human concerns and beliefs about its nature—is the basic assumption that underlies the supernatural, religious worldview. We call it the "postulate of design."

The postulate of a purposefully designed universe, as we have seen, destroys any meaning we might hope to find in the experimental *method* of science. But in doing, it also ensures that it will never be incompatible with any of the *findings* of science.

This ability of the supernatural view to adjust itself to any finite set of facts has, ironically, made it seem easy to accept both the findings of science and the consolations of spiritualism. Scientists, as human beings, are susceptible to the temptation of these comforts. Some believe that the world of the supernatural lies just beyond where they are performing their controlled experiments, although they usually feel that it is even more evident in fields other than their own. However, we need not reject their results. As long as they are honest—reporting not only their conclusions but also their methods and reasoning—such nonmaterialist scientists can still contribute to the progress of science in their own fields of study.

Scientific Integrity

The issue at stake here is whether or not our worldview is to possess consistency and integrity. Science has worked so well and has been so successful that it is difficult, if not impossible, to live in the modern world while rejecting its findings. But by accepting those findings as a free bounty—while rejecting the hard assumptions and hard work that made them possible—the super-naturalist embraces a lie.

The Flagrant Lie

For the first time in history a civilization is trying to shape itself while clinging desperately to the animist tradition to justify its values, and at the same time abandoning it as the source of knowledge, of truth For their moral bases the "liberal" societies of the West still teach—or pay lip-service to—a disgusting farrago of Judeo-Christian religiosity, scientific progressism, belief in the "natural" rights of man, and utilitarian pragmatism. ... All these systems rooted in animism exist at odds with objective knowledge, face away from truth, and are strangers and fundamentally hostile to science, which they are pleased to make use of but for which they do not otherwise care. The divorce is so great, the lie so flagrant, that it afflicts and rends the conscience of anyone provided with some element of culture, a little intelligence, and spurred by that moral questioning which is the source of all creativity.-Jacques Monod, *Chance and Necessity*, 1971.

It is often claimed that science can say nothing about values and ethics because it can only tell us what is—not what ought to be. But once again this is a case of attempting to divorce the findings from the method of science. Properly understood, science tells us not only what is but also ***how we must behave if we are to understand what is***. Science has succeeded as a cooperative human effort by asserting the belief that the universe can only be understood through the values of integrity and truth-telling. In the process it has *become* a system of values, and it has provided humankind with a language which transcends cultural boundaries and connects us in a highly satisfying way to all the observable universe. It also has the potential to be used as the basis for a workable and profoundly satisfying system of ethics. Indeed, it *must* be so used if we are to accept its findings without self-deceit.

Overcoming Ignorance

A naturalistic system of ethics is not likely to be popular, however, until science can overcome the currently evident public attitude of ignorance and hostility. In response to a recent **San Diego Union** story outlining new developments in cosmological theory, a reader pointed out that "God is in control of the universe, and the sooner these so-called scientists realize this, they will not need to invent hocus-pocus 'dark or unseen matter' as a man-made explanation instead of acknowledging the true source of all things, the all-powerful, omnipotent, omnipresent God, the creator."

He's right, of course. Accept the supernatural and the hard work of making and testing theories becomes a pointless enterprise, along with all human-made explanations and meaning. But if we allow such myths to limit the scope and uses of science, we will do so to our own peril and shame.

What Started the Clock

In an article in the October 4, 1985, issue of *Science,* cosmologist Steven Weinberg said that, even if science manages to trace the materialist explanation back to the first ten-billionth of a second of the existence of the universe, we still don't know what started the clock. "It may be that we shall never know," he wrote, "just as we may never learn the ultimate laws of nature. But I wouldn't bet on it."

Thank you, Professor Weinberg. We needed that.

Science and Religion Are Interdependent

By Albert Einstein

"Science without religion is lame, religion without science is blind."

Science and Religion Are Interdependent

Albert Einstein's theory of relativity revolutionized physics and earned him the Nobel Prize in 1921. Einstein renounced his German citizenship in 1933, when Adolf Hitler gained power. That year he accepted a position at the Institute for Advanced Study in Princeton, New Jersey and lived there until his death in 1955. The following viewpoint is an essay Einstein wrote in 1941 in which he contends that science and religion depend upon each other. While disagreeing with the idea of a personal God who controls the universe, he maintains that great scientists are inspired by a religious faith that the universe is rational.

As you read, consider the following questions:

1. Why does Einstein argue that ideally, there should be no conflicts between religion and science?
2. How does religion influence and benefit scientists, according to the author?
3. What harmful consequences does Einstein believe result from belief in a personal God?

It would not be difficult to come to an agreement as to what we understand by science. Science is the century-old endeavor to bring together by means of systematic thought the perceptible phenomena of this world into as thorough-going an association as possible. To put it boldly it is the attempt at the posterior reconstruction of existence by the process of conceptualization. But when asking myself what religion is I cannot think of the answer so easily And even after finding an answer which may satisfy me at this particular moment I still remain convinced that I can never under any circumstances bring together, even to a slight extent, all those who have given this question serious consideration.

The Definition of Religion

At first, then, instead of asking what religion is I should prefer to ask what characterizes the aspirations of a person who gives me the impression of being religious: A person who is religiously enlightened appears to me to be one who has, to the best of his ability, liberated himself from the fetters of his selfish desires and is preoccupied with thoughts, feelings, and aspirations to which he clings because of their super-personal value. It seems to me that what is important is the force of this super-personal content and the depth of the conviction concerning its overpowering meaningfulness, regardless of whether any attempt is made to unite this content with a divine Being, for otherwise it would not be possible to count Buddha and Spinoza as religious personalities. Accordingly, a religious person is devout in the sense that he has no doubt of the significance and loftiness of those super-personal objects and goals which neither require nor are capable of rational foundation. They exist with the same necessity and matter-of- factness as he himself. In this sense religion is the age-old endeavor of mankind to become clearly and completely conscious of these values and goals and constantly to strengthen and extend their effect. If one conceives of religion and science according to these definitions then a conflict between them appears impossible. For science can only ascertain what *is,* but not what *should be,* and outside of its domain value judgments of all kinds remain necessary. Religion, on the other hand, deals only with evaluations of human thought and action: it cannot justifiably speak of facts and relationships between facts. According to this interpretation the well-known conflicts between religion and science in the past must all be ascribed to a misapprehension of the situation which has been described.

For example, a conflict arises when a religious community insists on the absolute truthfulness of all statements recorded in the Bible. This means an intervention on the part of religion into the sphere of science; this is where the struggle of the Church against the doctrines of Galileo and Darwin belongs. On the other hand, representatives of science have often made an attempt to arrive at fundamental judgments with respect to values and ends on the basis of scientific method, and in this way have set themselves in opposition to religion. These conflicts have all sprung from fatal errors.

Now, even though the realms of religion and science in themselves are clearly marked off from each other, nevertheless there exist between the two strong reciprocal relationships and dependencies. Though

religion may be that which determines the goal, it has, nevertheless, learned from science, in the broadest sense, what means will contribute to the attainment of the goals it has set up. But science can only be created by those who are thoroughly imbued with the aspiration towards truth and understanding. This source of feeling, however, springs from the sphere of religion. To this there also belongs the faith in the possibility that the regulations valid for the world of existence are rational, that is, comprehensible to reason. I cannot conceive of a genuine scientist without that profound faith. The situation may be expressed by an image: Science without religion is lame, religion without science is blind.

A Strong and Noble Force

The ethical behavior of man is better based on sympathy, education and social relationships, and requires no support from religion. Man's plight would, indeed, be sad if he had to be kept in order through fear of punishment and hope of rewards after death.

It is, therefore, quite natural that the churches have always fought against science and have persecuted its supporters. But, on the other hand, I assert that the cosmic religious experience is the strongest and the noblest driving force behind scientific research. No one who does not appreciate the terrific exertions, and, above all, the devotion without which pioneer creations in scientific thought cannot come into being, can judge the strength of the feeling out of which alone such work, turned away as it is from immediate practical life, can grow. What a deep faith in the rationality of the structure of the world and what a longing to understand even a small glimpse of the reason revealed in the world there must have been in Kepler and Newton to enable them to unravel the mechanism of the heavens in long years of lonely work!

Though I have asserted above that in truth a legitimate conflict between religion and science cannot exist I must nevertheless qualify this assertion once again on an essential point, with reference to the actual content of historical religions. This qualification has to do with the concept of God. During the youthful period of mankind's spiritual evolution human fantasy created gods in man's own image, who, by the operations of their will were supposed to determine, or at any rate to influence the phenomenal world. Man sought to alter the disposition of these gods in his own favor by means of magic and prayer. The idea of God in the religions taught at present is a sublimation of that old conception of the gods. Its anthropomorphic character is shown, for instance, by the fact that men appeal to the Divine Being in prayers and plead for the fulfillment of their wishes.

Nobody, certainly will deny that the idea of the existence of an omnipotent, just and omnibeneficent personal God is able to accord man solace, help, and guidance; also, by virtue of its simplicity it is accessible to the most undeveloped mind. But, on the other hand, there are decisive weaknesses attached to this idea in itself, which have been painfully felt since the beginning of history. That is, if this being is omnipotent then every occurrence, including every human action, every human thought, and every human feeling and aspiration is also His work; how is it possible to think of holding men responsible for their deeds and thoughts before such an almighty Being? In giving out punishment and rewards He would to a certain extent be passing judgment on Himself. How can this be combined with the goodness and righteousness ascribed to Him?

The Source of Conflicts

The main source of the present-day conflicts between the spheres of religion and of science lies in this concept of a personal God. It is the aim of science to establish general rules which determine the reciprocal connection of objects and events in time and space. For these rules, or laws of nature, absolutely

general validity is required—not proven. It is mainly a program, and faith in the possibility of its accomplishment in principle is only founded on partial successes. But hardly anyone could be found who would deny these partial successes and ascribe them to human self- deception. The fact that on the basis of such laws we are able to predict the temporal behavior of phenomena in certain domains with great precision and certainty is deeply embedded in the consciousness of the modern man, even though he may have grasped very little of the contents of those laws. He need only consider that the planetary courses within the solar system may be calculated in advance with great exactitude on the basis of a limited number of simple laws. In a similar way, though not with the same precision, it is possible to calculate in advance the mode of operation of an electric motor, a transmission system, or of a wireless apparatus, even when dealing with a novel development.

To be sure, when the number of factors coming into play in a phenomenological complex is too large, scientific method in most cases fails us. One need only think of the weather, in which case prediction even for a few days ahead is impossible. Nevertheless no one doubts that we are confronted with a causal connection whose causal components are in the main known to us. Occurrences in this domain are beyond the reach of exact prediction because of the variety of factors in operation, not because of any lack of order in nature.

Rapt in Awe

The most beautiful thing we can experience is the mysterious. It is the source of all true art and science. He to whom this emotion is a stranger, who can no longer pause to wonder and stand rapt in awe, is as good as dead: his eyes are closed. This insight into the mystery of life, coupled though it be with fear, has also given rise to religion. To know that what is impenetrable to us really exists, manifesting itself as the highest wisdom and the most radiant beauty which our dull faculties can comprehend only in their most primitive forms—this knowledge, this feeling, is at the center of true religiousness. In this sense, and in this sense only, I belong in the ranks of devoutly religious men.

We have penetrated far less deeply into the regularities obtaining within the realm of living things, but deeply enough nevertheless to sense at least the rule of fixed necessity. One need only think of the systematic order in heredity, and in the effect of poisons, as for instance alcohol, on the behavior of organic beings. What is still lacking here is a grasp of connections of profound generality, but not a knowledge of order in itself.

An Ordered, Regular Universe

The more a man is imbued with the ordered regularity of all events the firmer becomes his conviction that there is no room left by the side of this ordered regularity for causes of a different nature. For him neither the rule of human or the rule of divine will exists as an independent cause of natural events. To be sure, the doctine of a personal God interfering with natural events could never be *refuted*, in the real sense, by science, for this doctrine can always take refuge in those domains in which scientific knowledge has not yet been able to set foot.

But I am persuaded that such behavior on the part of the representatives of religion would not only be unworthy but also fatal. For a doctrine which is able to maintain itself not in clear light but only in the dark, will of necessity lose its effect on mankind, with incalculable harm to human progress. In their struggle for the ethical good, teachers of religion must have the stature to give up the doctrine of a personal God, that is, give up that source of fear and hope which in the past placed such vast power in the hands of priests. In their labors they will have to avail themselves of those forces which are capable of cultivating the Good, the

True, and the Beautiful in humanity itself. This is, to be sure, a more difficult but an incomparably more worthy task. After religious teachers accomplish the refining process indicated they will surely recognize with joy that true religion has been ennobled and made more profound by scientific knowledge.

Spirituality's Role in Science

If it is one of the goals of religion to liberate mankind as far as possible from the bondage of egocentric cravings, desires, and fears, scientific reasoning can aid religion in yet another sense. Although it is true that it is the goal of science to discover rules which permit the association and foretelling of facts, this is not its only aim. It also seeks to reduce the connections discovered to the smallest possible number of mutually independent conceptual elements. It is in this striving after the rational unification of the manifold that it encounters its greatest successes, even though it is precisely this attempt which causes it to run the greatest risk of falling a prey to illusions. But whoever has undergone the intense experience of successful advances made in this domain, is moved by profound reverence for the rationality made manifest in existence. By way of the understanding he achieves a far-reaching emancipation from the shackles of personal hopes and desires, and thereby attains that humble attitude of mind towards the grandeur of reason incarnate in existence, and which, in its profoundest depths, is inaccessible to man. This attitude, however, appears to me to be religious, in the highest sense of the word. And so it seems to me that science not only purifies the religious impulse of the dross of its anthropomorphism but also contributes to a religious spiritualization of our understanding of life.

Rational Knowledge

The further the spiritual evolution of mankind advances the more certain it seems to me that the path to genuine religiosity does not lie through the fear of life, and the fear of death, and blind faith, but through striving after rational knowledge. In this sense I believe that the priest must become a teacher if he wishes to do justice to his lofty educational mission.

Science and Religion Are Not Enemies

By John Oakes

Many words have been written and much rhetoric produced coming from scientific materialists and "New Atheists" such as Richard Dawkins, declaring that the human religious pursuit is the natural enemy of human progress, and more particularly, of the free search by scientists for knowledge about the physical world. Famously, Stephen Jay Gould and Niles Eldridge have called for peace between warring scientists and religionists by declaring that science and religion are nonoverlapping magisteria. According to Gould and others, there is no overlap in subject matter or in the kinds of questions to be asked and answered by the purveyors of religion and science; therefore, the two can simply ignore one another. It is not hard to read between the lines of Eldridge's words to detect his assumption that, with time, the human need for religion, reflecting a premodern superstition, will soon conveniently disappear.

The question at hand here is this: What is the relationship between science and religion as they are regularly practiced in modern life? Is their language and means of acquiring knowledge incommensurate? Are there any important questions that both religion and science seek to answer? And if so, might their means of addressing these questions be complementary, rather than being in a natural and unending state of conflict? Does the arrival of the age of science herald the inevitable decline and fall of religion? What kinds of questions is science good at answering, and what are the limits of science? The same should be asked of religion. What is it good at doing, and what are its limits—areas in which it generally is not productive?

One conclusion will be that, although science and religion are broadly incommensurate, there are areas of inquiry where they overlap. The other will be that it is a mistake to assume that the two are natural enemies. Scientific inquiry is not the natural enemy of religious pursuit. Neither is religion, if pursued in its appropriate context, the natural enemy of the scientific search for knowledge about the universe.

What Is Science, and What Are Its Limits?

First, of course, we need to know what we are talking about. What is science? Putting aside for a moment the claim of philosophers that there is no real philosophically defendable scientific method, what is science, and what can we learn from the scientific approach to acquiring knowledge? Put simply, science is a means to discover the underlying laws that govern the natural world, using empirically generated data, as well as theories and models to explain that data. Science does not answer the ultimate question "Why?" Rather, science provides us with explanations of physical phenomena that are not self-contradictory and that are consistent with the physical evidence. Science provides us with physical explanations of physical phenomena.

By its very nature, science is limited in the kinds of knowledge it can give us. It is very good at answering certain questions and very bad at answering others. Its conclusions are always tentative and never the final answer. For this reason, science does not answer the deeper questions about truth. It is completely unable to answer the metaphysical question: Why? On the other hand, science is really quite effective in answering questions such as Where? When? How many? By what means? Arguably, it is the by far the most effective means yet devised by human beings to answer such questions. Postmodernists may question whether absolute truth exists, but science certainly does seem to give extremely reliable knowledge about the workings of the physical world.

Furthermore, science is quite limited and perhaps even useless to answer questions such as "What is the value of human life?" "Is that the right thing to do?" "Am I here for a reason?" Without exception, human beings find themselves asking questions about beauty, social justice, and purpose. Like a local news commentator here in San Diego says when government officials ignore the needs of regular people, "It ain't right!" Science does not help us here. In assessing the relative importance and need for science in human societies, it is worth noting that these are the kinds of questions people really care about. Human beings are not as concerned with where, when, and how many, but are very concerned with questions of justice and truth. When I discuss the limits of science with my students, I point out that in the final analysis, science is not very good at answering any of the questions most of us really care about. This is not to deny the importance and usefulness of science. Through science we have cured diseases, understood the marvelous workings of nature on a microscopic and cosmological level, and are able to predict our future and devise means to avoid the negative consequences of human behavior. However, it is clear that science is not the only means for asking and answering questions, and its ability to answer the questions humans care most deeply about is limited. In order to meet the needs of real people and to maximize the human good, other sources of knowledge and experience, such as art, philosophy—and perhaps religion—are essential.

What Is Religion, and What Are Its Limits?

It is clearly difficult to define religion, and even more difficult to assess its limitations. However, we must make the attempt in order to assess if religion and science are natural opponents. Scientists generally agree, at least broadly, on a "method" to acquire knowledge of the world. Clearly, humans do not agree on the "right" religion. Yet, we can establish in very broad outline the sphere of knowledge and the means of establishing that knowledge in the human activity we label as religion. Generally, religion asks questions such as the place of human beings in the world—not just the physical world, but in the larger world, which includes purpose and meta realities which may or may not exist outside and above physical things. Those who practice religion ask questions of what is right and wrong. They ask not what is, but what ought to be. What is my purpose? Is there a higher, supernatural reality? If so, what is the human relationship to that reality? Whereas science seeks tentative explanation and rejects authority, religion, at least in this sense, is the opposite. For the most part, religious "truth" and knowledge are based on authority, such as that of a guru or a canonical scripture. In science, nothing is true, per se, but in most religious contexts, truth is well defined. Scientific knowledge changes and grows. Religious experience may change and grow, but religious claims do not.

We may be stepping into controversial territory here, but generally, religion is not particularly effective in answering questions about measurable things. Questions such as when, where, how many, and so forth are either not answered, or the track record for religions answering such questions has not held up all that well. We ignore the history of this to our peril. It seems not unreasonable to conclude that generally, religion can concede to science the role of informing us the cause of a particular disease, the history of the universe, the age of rock formations, and the probable result of combining certain chemicals.

Boundaries

Humans are social beings, but we are individuals as well. In a social sphere, usually we will concede space to the other, but in our own personal sphere, we will defend our own territory vigorously. I will share space with my neighbor at the coffee shop, but will not concede space to him or her in my own bed. The general conclusion from the discussion above is that the "homes" of science and religion are separate. These are more or less incommensurate bodies of knowledge. As long as religion does not enter the bedroom of science and science does not enter the bedroom of religion, we can have peace. It should not surprise us that when religion invades the natural territory of science, it evokes a reaction, and vice versa. If science tries to declare that alcoholism is neither right nor wrong, religion will not concede this point. If religion tries to declare that "sin" is the cause of disease, science will not remain silent. Nor should it.

If scientific materialists try to tell us that, based on experiments in neuroscience, the human soul and human consciousness are not real, or at best epiphenomena, then it seems fair for those with religious faith to cry foul. Since when could science answer questions about ultimate reality? This is a boundary issue. Scientists would be best to take off their scientist hats before speaking on such a topic they know little, if anything, about. Unfortunately, some scientists do not respect this boundary.

On the other hand, if a person with faith in a particular religious authority declares that their scripture denies that the earth moves or claims that the universe has existed in an infinite cycle—a wheel of time—then the scientist has reason to cry foul as well. If a religious claim tells us that galaxies do not exist, the scientist seems within his or her right to respond that this religious claim is almost certainly not true. Again, this is a boundary issue. At the very least, the person with religious faith ought to hesitate to impose a qualitative belief on quantitative science.

Perhaps humility might go a long way here. The scientist ought to hesitate to declare that the physical world is all there is—that there is no God, no supernatural reality—and the person of faith ought to pause before declaring a particular scientific conclusion to be false doctrine. Is it not possible that their own interpretation of their authority is what is at fault? Or as St. Augustine proposed, such an anomaly may be evidence, not that science is wrong, but that their religious authority might be mistaken. The story of Galileo's conflict with the Roman Curia is informative here. On the one hand, for the materialist to declare, by fiat, that there is no supernatural intervention in the world is to commit a boundary error. On the other hand, for a person of faith to apply such a faith to declare that there are no truly random forces in nature seems to be a boundary error as well.

When Do Science and Religion Overlap, and How Should This Be Handled?

It would be nice if life were simple. One can only wish that Gould and Eldridge are completely right that science and religion are nonoverlapping. However, reality is complex. The fact is that there is indeed overlap between the territories of science and of religion. Is human consciousness real or a mere epiphenomenon? Is there a real demarcation between humans and other animals? If so, what is that demarcation? Was the physical universe created? If so, how and why? Was life created, and can fully random forces explain the creation of life? Given the apparent "phase transition" of complexity between living and nonliving things, might there be a corresponding transition to a higher level of reality? Is a religious experience just chemicals moving around in our brain, or might such chemical activity be an indicator of something real happening on another level of reality? Is love just the release of neurotransmitters and the firing of certain neurons, or might "love" be something real? Do I exist? Do I have a body, or am I a body? Neither science nor religion has exclusive ownership of any of these questions. It is in these areas that each can inform the other and that, for the wise person, such interchange will indeed happen.

To simply declare that religion has nothing to offer to these questions or that such questions are sheer nonsense is not acceptable to the great majority of people. To do so is to undermine the dignity of human beings and to lessen the value and quality of life. On what authority can anyone declare such questions nonsense? To say that justice is a meaningless word and that religious experience is mere superstition is to declare the result of an experiment that has not even been performed.

On the other hand, for persons with religious faith to ignore the implications of genetic research into the causes of alcoholism or the discoveries of neuroscience is shortsighted. Perhaps one can even argue that the moral imperative of most religions includes the search for truth, wherever it leads. One can argue that to reject on religious presuppositional grounds the implications of scientific discoveries is to lessen the value and quality of life as well. If it is foolish to simply declare religious experience foolishness, it is also foolish to simply ignore the vast and growing evidence for the common descent of life on the earth.

Conclusion: Science and Religion Should Be Friends

The conclusion to this point is that on a great number of questions, science and religion are incommensurate. Careful attention to boundaries can, for the most part, allow the two to coexist without doing battle. Humility and caution can allow people to delve into the areas where the two overlap without major friction. Science and religion can coexist in peace. However, the conclusion of this essay is not just that the two can exist in peace. The claim is that they are natural friends. Is this going too far? Let me explain.

Let us consider the question of alcoholism. If we only listen to the "science," perhaps we will notice the genetic predisposition of some to alcoholism, but fail to give hope to the alcoholic. It is not inconceivable that if we do not allow science and religion to work together, we may leave the alcoholic in a very bad place. The science alone might even give the person an excuse to not change. Perhaps the "ought" of religion can make the difference for a person to overcome the addiction. On the other hand, if we only consider the "religion" of alcoholism, declaring it a sin but ignoring the science, we may miss a chance to use a chemical treatment to help the person overcome alcoholism. We might also fail to show compassion, not understanding that for some it really is harder than for others, for reasons not completely within their control.

Does understanding the brain chemistry of prayer make it any less benefit to the believer who prays? Perhaps knowing that her brain was "designed" to allow her to experience both a spiritual and a physical effect from prayer might increase the faith of a believer. Many believing scientists have found special revelation from religion and general revelation from science to complement one another. Galileo had a good grasp of the boundary issues and the complementary nature of science and religion. In his letter to the Duchess Christina (1614), speaking of his Christian religion and science he said, "I think that in discussions of physical problems we ought to begin, not from the authority of scriptural passages, but from sense-experiences and necessary demonstrations; for the Holy Bible and phenomena of nature proceed alike from the divine Word, the former as the dictate of the Holy Spirit and the latter as the observant executor of God's commands." If we allow science and religion to work together, especially in that limited number of questions on which they naturally overlap, much good can result. We can contemplate not just the truth that God (or the gods, or Brahman or…) created all, but marvel at how it was done. If we allow for the possibility of a design or a plan, then a vast array of incoherent but amazing discoveries can become coherent. They will make more sense. If we respect boundaries, how is science hindered by religion? The answer, historically, is that religion will inform science. That certainly was the case with Roger Bacon, Copernicus, Galileo, and all the early scientists. The answer is that if we respect boundaries, science will inform religion as well. If we can assume that our scripture or religious authority is a source of real truth, then science might even help us to understand how to interpret revealed truth. As one believer has said, all truth is God's truth.

In summary, science and religion are natural friends. If those who practice science and religion will respect reasonable boundaries, allow humility and reason to prevail in the places where the two overlap, and if they will be informed by science and religion when *both* are relevant to important questions, then science and religion can be kissing cousins once again.

CPSIA information can be obtained
at www.ICGtesting.com
Printed in the USA
BVHW060116070220
571513BV00003B/171